Moving Wearables into the Mainstream

Taming the Borg

Moving Wearables into the Mainstream

Taming the Borg

by

Joseph L. Dvorak
Motorola
USA

Joseph Dvorak

joseph.dvorak@gmail.com

ISBN-13: 978-0-387-69139-8 e-ISBN-13: 978-0-387-69142-8

Library of Congress Control Number: 2007934286

Printed on acid-free paper.

9 8 7 6 5 4 3 2 1

springer.com

To my wife Barbara and children Katie and Justin

Acknowledgments

This book and its material was a true collaborative effort. First, the book itself could never have been completed without the love and support of my wife Barbara and my children Katie and Justin. Their patience and belief in my ability to write the book were invaluable.

Much of the design principles and processes discussed in the book came from two major wearables projects at Motorola. The first one, Person Integrated Communications (PIC), was unfailingly supported by my manager at the time Darrell McClendon. I am truly grateful for his firm belief in the project and his efforts to keep the project going. I would also like to thank Scott Greven who did much of the work on the communications systems for the project. His excellent work and attention to detail made short work of some of the most challenging parts of the prototype systems.

The principle sponsor and supporter of the follow-on project, Conformables, was my director in iDEN Technology at Motorola, Jaime Borras. He continually urged us on in the face of many who doubted. We would not have gotten as far as we did without his support and I thank him.

Much of the prototyping and design for Conformables was done by Luke Lenzen, a three year intern who was the other major co-developer on the project. His ideas and enthusiasm were major factors in our success. I would also like to thank Alan Beatty and David Hayes who provided hardware and systems level support for the prototypes.

Ryan Nilsen had a major impact on the formulation of the design principles in this book and the design language adopted for many of the prototypes. His designs were always very functional and very elegant and were a major reason for the excitement these prototypes generated in all who saw them.

Philip Hodgson, Thad Starner, and Sandy Pentland all provided valuable sources and comments during their review of the book. I was also fortunate to have Philip lending human factors advice during the PIC project. Randy Vaas from Motorola provided valuable advice on legal aspects of wearables in social contexts.

I am also deeply grateful to my wife who painstakingly proofread the manuscript and caught numerous mistakes.

Finally, I would like to thank Susan Lagerstrom-Fife, Sharon Palleschi, and the other professionals from Springer-US for their patience, advice, and support during this project.

Preface

We are awash in technology. It is in our homes, our workplace, and our cars. It is almost impossible to escape its influence on our lives. Our technology demands our attention. Cell phones interrupt us regardless of what we are doing or where we are. PCs frustrate us and make us feel technologically inadequate.

And there is more on the way. Technologies such as biosensors, augmented reality, and pervasive computing promise to literally immerse us in a sea of technology, all meant to make our lives easier.

One of these promising technologies is wearables. Wearables have received much attention in the past. The image usually shows someone carrying a portable computer in their clothing or on their belt.

However, this is wearable only a geek could love. Graduate students and researchers are comfortable with carrying and using large, obtrusive and complex devices. Their focus is on pushing the envelope of current technology. Ease of use and comfort are usually a lower priority.

Current wearable technology has also been adopted in specialized application areas such as vehicle maintenance, inventory control, and the military. These applications involve sophisticated tasks for which the availability of computing power and special applications in mobile environments outweigh the current levels of obtrusiveness and application complexity.

The term 'wearables' encompasses a wide spectrum of devices, services, and systems Objects from entire desktop equivalent computers to a ring with an RFID chip have been referred to as wearables. Not all of these types of wearables will be accepted by the mainstream. Many wearable form factors will have to change significantly if they are to be widely adopted by consumers. The obtrusiveness of the devices, the complexity of the applications, and the geeky appearance must be reduced. For the most part, users of current wearables must conform to the constraints and limitations of the technology and applications. While this may be acceptable in specialized application areas, the general user population will not accept it. For these users, the technology and applications must conform to them.

This book discusses the characteristics and design elements required for wearable devices and systems to be widely adopted by the mainstream population for use in their everyday lives. We introduce concepts such as Operational Inertia that form a mindset conducive to designing wearables suitable for adoption by the mainstream.

But there is more to designing a wearable than selecting the appropriate technology and form factors. Technology is not used in a vacuum. This is especially true for wearables. Since wearables are by their nature closely associated with the person, their use generates many social and even legal issues that have little to do with specific technologies. We discuss the implications of these issues for mainstream wearable systems since it is these issues that can pose the greatest impediment to their successful adoption.

Wearable technology has actually been with us throughout history. An early example is the development of buttons providing an easy method for keeping shirts and jackets closed. More recent examples include the development of synthetic materials such as nylon and rayon. However, it has been so successfully integrated into people's daily lives that those in the past never regarded it as a specific technology. It was only when wearable computers started to appear and their form factors were highly incompatible with easy wearability and the concepts of fashion that wearables as a specific technology gained real visibility.

With the new generation of PDAs and smartphones, the term wearable is now in flux. There is no agreed upon definition of wearable. One of the most comprehensive definitions is the set of characteristics and attributes of a wearable computer defined by Steve Mann in 1998. This definition needs to be refined for wearables aimed at the mainstream population. The characteristics of not monopolizing the user's attention and not restricting the user's activities need additional emphasis and development. The primary vehicle for the refinement and increased emphasis in this book is Operational Inertia. Operational Inertia is defined as the resistance a device, service, or system imposes against its use due to its design. The minimization of Operational Inertia is a constant design theme.

Most current applications of wearables are for wearable computers. These are primarily focused on specialized activities such as vehicle maintenance and military operations. An area with broader potential appeal is wellness maintenance where the wearable monitors the user's health and activities and presents information and suggestions to maintain and improve it. While significant design challenges remain, wellness maintenance could be one of the most useful applications for the mainstream population. Other areas of promising applications are cognitive assistance and personal security.

Wearables will provide their most widespread benefits in the seamless integration of people's everyday tasks. By being aware of the user's environment and activities the wearable system will transparently and, in many cases, proactively assist the user with whatever they are doing. We will increase our effectiveness. But more importantly, we will increase our ability to concentrate on those things that are really important to us.

To achieve these benefits wearables will incorporate a wide array of technologies. Many of these technologies, such as graphical user interfaces, are mature while others, such as speech recognition, are starting to find their way into commercial products. Many others are still in the research or early development stage. Examples include data fusion for context awareness, indoor location systems, smart fabrics and clothing, and activity detection and reasoning.

Combining these technologies into a system that is powerful but easy to use and unobtrusive presents significant design challenges. Several factors will determine how readily the mainstream population accepts wearables. Among these, wearability, ease of use, and compelling form factors are most relevant to design. Often these factors cannot all be optimized simultaneously so tradeoffs must be made.

But the technical issues, as significant as they are, may not be the most challenging. The use of wearables will generate many social issues as well. Issues of privacy, violation of social conventions, dependency on technology, and others will arise as people utilize wearables within their daily tasks and social interactions. Society is starting to discuss some of these issues now with the increasing use of camera phones for example. The intimate association of the wearable with the person will give new urgency and scope to these issues.

As is often the case with technology, laws and the legal system will play catch up. Issues of personal responsibility for actions by intelligent and mostly autonomous software agents, where the person ends and the wearable begins relative to police searches, and the use of these devices in testing situations or sensitive areas will be argued in and deliberated by the courts for some time.

Regardless of the social issues and legal landscape that evolves, wearables in some form will be used. Their full potential is hard to imagine. New technologies such as flexible displays, brain – computer interfaces, and totally implantable devices will completely alter wearables as we envision them today. The incorporation of neural networks, emotion and personalities, and commonsense reasoning will provide us with wearable systems with unimagined power and intelligence. This will only further strengthen the relationship between the user and their wearable system. It may be no exaggeration to say that the relationship a user develops with their wearable system will become highly symbiotic, intimate, and could usher in a new view of what being human really means.

Intended audience

This book can be used as a textbook in an introductory course on wearable technology. It provides a broad discussion of the various technologies underlying wearable devices and how those devices could be designed for acceptance by the mainstream population. Current practitioners of wearable research will find it useful in giving them an overview of the many areas outside their own that are relevant to wearables and that may form the context in which they conduct their research. Finally, this book will prove useful to anyone interested in a broad overview of wearable technology.

This book strives to answer the question: how can we design wearable devices, services, and systems that ordinary people can use to help them with their daily tasks without having those wearables getting between the user and the tasks they are trying to do? In short, how do we design wearables that can be used transparently? The answers to these questions will, in large part, define the level of acceptance of wearables by the mainstream population.

Joe Dvorak

Contents

PART 1: INTRODUCTION TO WEARABLES

Chapter 1

BACKGROUND

1.1 POWER TO THE PEOPLE

In the coming decades we will witness an extraordinary change in how we focus on and interact with technology. As shown in Figure 1-1, technology in the 1990s reflected the power of the microprocessor. Moore's Law was in full force and processing power in computers was increasing rapidly.

The first decade of the 21st century reflects the power of the network, specifically the Internet. The capabilities of the Internet have spawned whole new areas of applications. Web sites such as MySpace, Facebook, and Second Life are part of the rise of social networking and virtual worlds.

In the second decade of the 21st century technology will reflect the power of people [1]. Technology will enable people to compensate for missing or impaired capabilities to a degree unheard of today. Technology will also augment and enhance our existing capabilities far beyond what we now consider normal. An example of the former is the development of intelligent prosthesis that attach directly to the remaining part of an amputated limb and whose circuits interface to and communicate with the limb's nerves. An example of the latter is exoskeletons that attach to a person's body and greatly amplify the person's running and carrying abilities without impairing the natural movement of the arms and legs.

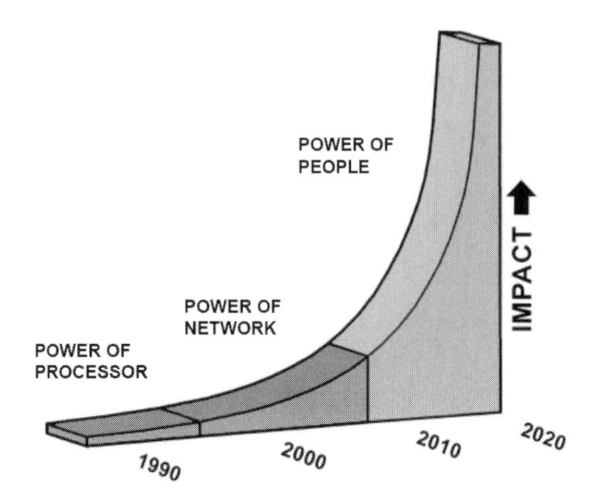

Fig. 1-1. The Power Curve *(MIT Media Lab)*

Wearables (of which the devices discussed above are the most extreme examples) are a natural technology to thrive in this focus on the power of people. Of all technologies wearables have the most intimate connection with people. They are worn by the person, are with them for prolonged periods of time, and, through supporting technologies such as context awareness, interface most closely and effectively with the person.

It is this intimate connection and interfacing with us that makes wearables such an important technology. No other technology has as much potential to monitor our well being, anticipate our needs, and assist us with our everyday tasks, regardless of where we are or what we are doing.

Surely the era of power of people should be a golden age for wearables.

1.2 A CURIOUS SITUATION

However, wearables is currently a technology in search of acceptance by the mainstream population. In 2002 VDC forecasted that global shipments of wearable computers will likely reach over $100 million in 2002 and grow to over $563 million in 2006 [2]. Most of this growth would have come from wearable computers in vertical markets and applications. In 2005 VDC drastically revised their forecasts downward, estimating the global market for general-purpose computing/communications wearable systems at $170 million in 2005 and $270 million by 2007 [3].

Clearly, the market forecasts for wearables (mostly wearable computers) over the last 10 years have been consistently way off. The market has not materialized as these forecasts predicted. With the exception of niche areas such as the military and specialized maintenance applications, wearables have not achieved wide acceptance. Many of the companies making wearable computers have gone out of business, been bought, or moved to another line of product.

The struggling nature of the technology can most clearly be seen in the viability of companies serving this market. Time has not been kind to most of these companies. Via has been acquired by InfoLogix [5], Charmed Technologies no longer sells wearable computers[1], and the commercial leader in wearable computers, Xybernaut [6], has yet to make a profit.

Why is this? Clearly, people are comfortable using portable, mobile devices - just look at the success of cell phones and PDAs. Well, for one thing, current wearables are systems only a geek could love or want to use. Most current wearable systems are obtrusive, unattractive, and complicated to put on and use. This is neither a surprise nor a criticism since most wearable systems are wearable computers that are research vehicles and/or aimed at specific activities such as vehicle maintenance, warehouse order fulfillment, or the military. In most of these cases users have little choice in wearing them so obtrusiveness and appearance take a back seat to functionality.

[1] Charmed Technologies has redirected its efforts toward the CharmBadge, a conference badge about the size of a business card that contains a small processor, memory, and IR transceiver [5].

If wearables are to be pervasively adopted by the mainstream user population, they must be nearly transparent to use. That is, they must aid the user in the performance of the user's primary task without bringing attention to themselves. In subsequent chapters we will discuss in detail the requirements for transparent use design.

Transparent use does not mean invisible. These wearables can still be highly visible, attractive, and enjoyable to use. However, the pleasure will be in using the device to increase the ease with which we get our everyday tasks done. In other words, we will appreciate the transparent assistance, not the functional attributes of the devices themselves. This illustrates the paradox of transparent use design: by making the device transparent to use, by making the technology invisible, the device can be better appreciated for its functionality.

Transparent use is important for another reason. As a term, 'wearables' embodies the incongruous combination of clothing and electronics. For many people wearable technology seems incompatible with fashion. Wearable technology is about chips, computers, circuit boards – all having the connotation of cold, logical, and devoid of feeling. Fashion, on the other hand, is all about self expression, comfort, and feeling. If the wearable system becomes transparent to use, the negative emotional connotations associated with the technology will not arise in the user's mind. This allows the user to concentrate on the positive feeling embodied by the clothing. This makes the augmented garment, and with it the wearable system, more acceptable to the mainstream population.

This book discusses what it takes to create a Mainstream Wearable System. A mainstream wearable system will succeed or fail based on the user experience it provides. The user experience must be one in which the user is minimally aware of the wearable system, allowing him to stay focused on his primary task and to complete it quickly. This requires some basic capabilities (Figure 1-2).

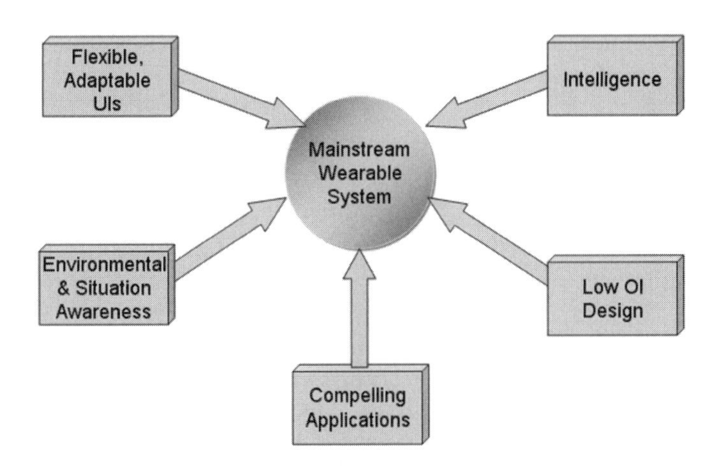

Fig. 1-2. Elements of a Mainstream Wearable System

Low Operational Inertia design creates devices, services, and systems that require very little setup effort, are very easy to use, and allow us to forget about them when they are worn but not used[2]. This requires us to completely rethink the form factor of a wearable system.

Environmental and situation awareness allows the mainstream wearable system to interact with us and the 'smart' devices within our environment using short range, lightweight communications. The system recognizes and acts upon the characteristics of our surroundings and situations. This enables the system to support our activities in the most effective manner.

Flexible, adaptable user interfaces means that we, the users, will be able to interact with our mainstream wearable system using whichever interface mechanism, or mechanisms, best support the current task and situation at

[2] Operational Inertia is presented in this book as a fundamental principle for developing wearables that are transparent to use. It is defined and discussed in detail in Chapters 4 and 5.

hand, whether that interface be graphical, text, speech, gesture, or one or more in combination.

Doing all of the above well requires that the system be intelligent – about us, our environment, and our social contexts. This intelligence requires more than simple rules and data. It requires a level of commonsense and reasoning, something basic to humans but a real challenge to incorporate into computers.

Finally, these attributes must be applied to the development of compelling applications. These applications will assist users in the performance of their everyday tasks without forcing them to focus on the applications themselves.

The first tentative steps in creating such mainstream wearable systems are being taken. There is a growing recognition that wearables must change into something more conducive to everyday use by the common person, in support of their everyday tasks. However, to understand the significance of this change it is helpful to trace the birth and early development of wearables. We can then better discuss how they are changing.

1.3 A BRIEF HISTORY OF WEARABLE TECHNOLOGY

One of the most used terms in the field of wearable computers is 'cyborg'. Early students and researchers in the field adopted the term to describe themselves and the kind of human – machine symbiot they thought would eventually evolve from the technology. The term was actually created by Manfred Clynes in 1960 [7]. The term was first used publicly at a NASA conference about human space exploration. At that conference 'cyborg' referred to an enhanced human that could survive in extraterrestrial environments. The researchers at the conference believed that such a man-machine hybrid would be needed in space flight and proposed a number of ways humans could be modified to survive in space. Clynes believed that the human and the spacecraft would have to be an interrelated system that shared information and energy. He created the term cyborg from cybernetic and organism, reflecting this relationship of interdependence.

An important figure in the earliest development of the field of wearable computers and one of the original cyborgs is Steve Mann. While still in high school Mann wired an eight bit 6502 computer into a steel-frame backpack to control flash bulbs, cameras, and other photographic systems. He designed and built the imaging system he wore to explore new concepts in imaging and lighting [8]. The display was a camera viewfinder CRT attached to a helmet. This provided a 40-column text overlay display. Input was from seven microswitches built into the handle of a powerful flash lamp, and lead-acid batteries powered the entire system (including flash-lamps). At that time battery-operated mobile computing was a totally new concept. In the 1980s there weren't any laptop computers, not to mention PDAs.

The 6502 microprocessor based computer was not powerful enough to do the desired image processing. Therefore, Mann developed a full duplex communication system between his wearable computer and a remote supercomputer. The link to the supercomputer was a high quality microwave link. The link back to his wearable was a lower quality UHF link. The processed image was received from the supercomputer and displayed on the head worn monocular display. With his system Mann explored such then novel concepts as mediated and augmented reality where text is overlaid upon scenes of the real world.

1.3.1 The Cyborg Era: 1990s

The 1990s saw significant growth in the area of wearable computers as well as several milestones. The first official wearable computer programs were established at universities. Thus began the age of serious research into the technology of wearables.

One the most well known wearable computer groups was at MIT. The group, headed by Sandy Pentland, developed many of the initial wearable applications and systems exploring context, fashion, and user modeling [9], [10]. Thad Starner and Steve Mann were graduate student members of this group. Like many other university based groups researching wearable computers, this group adopted the term 'cyborg' to define themselves in relation to their close interaction with the technology. Their gear was obtrusive, cumbersome, and strange looking to most non-cyborgs. And the cyborgs liked this sense of exclusivity and eccentricity that their appearance and behavior engendered.

Most early wearable computers were hand built since no companies were making kits or products in the early 90s. Researchers and students created their own hardware and software or adopted designs that were published by others in the cyborg community. These computers were severely limited by the technology at the time, especially the power generation technology. Most commercially available computers of the caliber these students were building were desktops attached to an AC outlet. Power consumption was not an issue for desktop PCs. But it was a major issue to the cyborgs since their computers had to be small and mobile.

The wearable computer field was hands on and had an air of a hobbyist culture. Individuals or small research groups built most of their wearable computers. There was often little documentation and user manuals were mostly unheard of. Most groups initially focused their attention on issues of infrastructure such as input devices, displays, and communications. Early areas of application focus included Computer Aided Cooperative Work, Augmented Reality, and Context Awareness. The emphasis was on the technology. Business and productization issues were of secondary importance.

Another early wearable computing group was at Carnegie Mellon [11]. Started in 1991, the group has developed prototypes of several wearable computers, aimed at a wide variety of areas including industry and military applications. Their research explored new ideas and resolve issues of wearability. The wearables, often produced at a rate of one design a year have ranged from designs involving systems integration on a task specification provided by a specific customer, to the more typical exploratory systems designed as pure research.

Over the years the group has developed several conceptual frameworks for wearable computers. The evolution of these frameworks, instantiated in their prototypes, forms a kind of "evolutionary tree" as shown in Figure 1-3 [12]. At the root of the tree are several supporting technologies such as miniature displays, speech recognition, microprocessors, language translation and wireless communications. As we travel up the tree, we see succeeding system implementations starting with Vu Man 1 in 1991. As

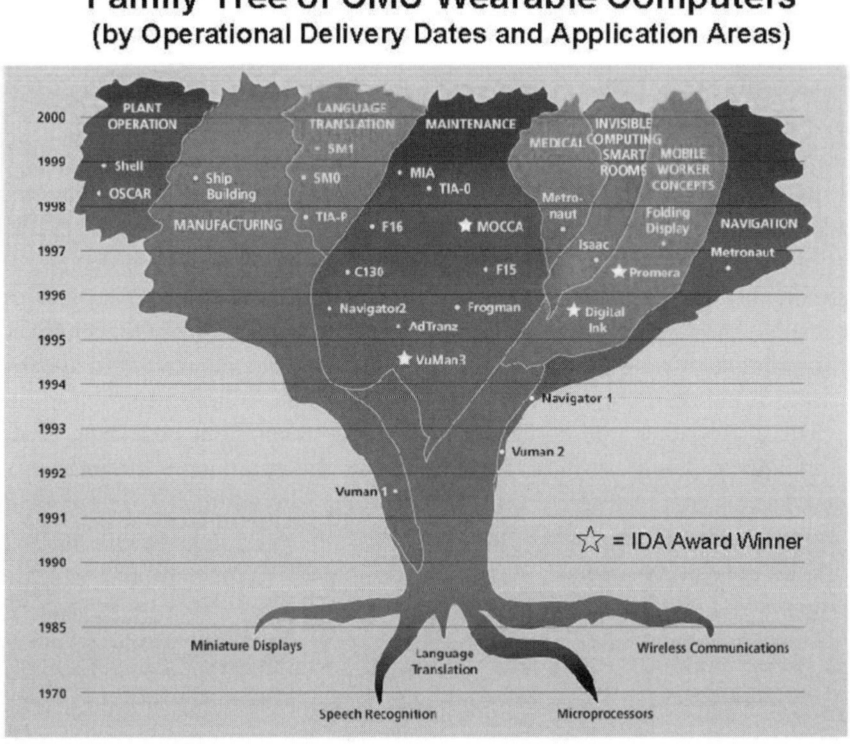

Fig. 1-3. CMU Wearable Computer Family Tree *(CMU Wearables Group)*

time passes, the various systems cluster into specific application areas, represented by areas at the top of the tree. These application areas include plant operation, manufacturing, language translation, maintenance, smart rooms, systems for mobile workers, and navigation.

In Europe, one of the most active groups is the Wearable Computing Lab at ETH Zurich (Swiss Federal Institute of Technology) [13]. Their research focuses on wearable architecture and devices. However, they also investigate supporting technologies such as conductive textiles, context awareness, and harvesting energy for wearable systems from thermal, optical and motion sources. This group has developed a prototype of what is at the time of this writing the smallest wearable computer, the QBIC Belt Integrated Computer discussed later in this chapter.

The growing research in wearable computers in the early and mid 1990s provided the impetus for the first organized international conference on wearable computers. The first International Symposium on Wearable Computing (ISWC) [14], was held in Cambridge, MA in 1997. It was sponsored by the IEEE Computer Society. Over the years it has become the most prestigious and well attended conference on wearable computers.

Early conferences had a very geeky feeling with few, if any, commercial products shown and many hand made system configurations. In the last couple of years however, the conference has taken on a more polished look as many groups started using general purpose portable computing platforms such as PDAs (a favorite is the HP iPAQ) and commercial products made by companies such as Xybernaut and Charmed Technologies started to appear.

There are other conferences on wearable computers and their related technologies. Chief among these are the International Conference on Pervasive Computing (Percom) [15] and the International Conference on Ubiquitous Computing (UbiComp) [16], both of which have strong wearable computing representation.

One of the most widely adopted wearable computer designs was the Lizzy [17]. The first Lizzy was developed in 1993 by Doug Platt and Thad Starner. The initial computer included the motherboard from a kit, a monocular display call the Private Eye, and the Twiddler, a one handed chording keyboard made by Handykey [18]. This system has been adopted and adapted by many researchers and students in the field.

The Lizzy design evolved as hardware improved. A version in the late 90s included:

- 150 MHz Pentium CPU
- 32 – 64 Mbytes of RAM
- 6 Gbyte hard disk
- Color VGA display driver
- 1 or 2 PCMCIA slots

- Cellular Digital Packet Data modem
- 1 or 2 camcorder batteries

While this configuration seems anemic, even by the standards of the desktops 5 years ago, it was cutting edge for wearable computers around 1998.

The 1990s also saw the start of commercial companies that manufactured wearable computer systems. Most of these companies aimed at specialized markets in industry. Of the early companies Xybernaut (founded in 1990) became the leading manufacturer of wearable systems. In 1999, Xybernaut released the Mobile Assistant (MA) IV. It contained a 200MHz Pentium MMX processor, 32 Mbytes of RAM and a 2.1-Gbyte hard drive. It supported Windows 95, 98 and NT as well as Linux. The system included a CPU unit with a belt holster, a head-mounted display over either eye that supported VGA and also contained a microphone and earphone. A wrist-mounted flat-panel touch-screen color display and keyboard were available as an option. Also included in the basic system was a battery pack and IBM Corp.'s ViaVoice speech-recognition software.

Another early wearable computer company was Via, Inc[3]. It produced the ViA II in 1999. It consisted of a unique segmented belt worn module that hugged the body. One of the segmented units contained a Cyrix 166 MHz processor with 64 Mbytes running Windows 95 or 98. The other segment contained the 6 Gbyte hard drive. The processor segment contained connectors for VGA output, audio input and output, and USB. A VGA touch screen tablet and heads up display were also available.

Charmed Technologies [5] was created in the late 1990s by former students from MIT. It was known mainly for its sponsorship of slick fashion shows called Brave New Unwired World. These shows, professionally

[3] Via was bought by Infologix [4].

staged and featuring top fashion models, showcased mostly highly fashionable conceptual mockups of wearable devices. Shows were held in conjunction with technology conferences such as Internet World in Berlin, London, Paris, Chicago and other cities. In 1999 Charmed introduced the 'CharmIt', a wearable computer based on the PC/104 architecture that was a standard among wearable computer designers. The CharmIt had a 266 MHz Pentium 2 processor with 64 Mbytes RAM. It also had VGA, USB, and serial ports as well as a PCMCIA slot and audio input and output jacks.

There was also another trend in the 90s: integrating computers into watches. Many people consider the watch as the ideal form factor for a wearable computer. Watches worn on the wrist are easily accessible and rarely get in our way. However, there are a number of problems embedding computation into a watch. The very characteristics that make a watch attractive as a wearable (small size, thinness, simple display) make them a poor choice as a host for significant computation. Wristwatches are inherently display only. The difficulties most people have setting the functions on the feature-rich digital watches only reinforce this point. There is very little room to display detailed information so help is usually not sufficient or not available on the watch itself. There is little room to place the type of controls that could be used to efficiently and easily enter and select information.

Nevertheless, from 1998 on there were many attempts at marrying the computer with the watch. The first one to gain any public following was the Seiko Instruments Ruputer [19]. The Ruputer is a mixture between PDA and a wristwatch (Figure 1-4 right). It connects to a PC via a serial cable and can be programmed with new functions. The Ruputer contained a 16-bit-CPU, running at 3 MHz. The display was monochrome with a resolution of 102x64 pixels, It contained 128Kbytes RAM and up to 4 Mbytes of flash memory. It most unique element was a joystick-like button that provided a random selection capability. It was powered by 2 Lithium coin cell batteries (CR 2025) that would last between 2 weeks and 3 months, depending on how often it was used. It came with several PDA like functions including Personal Information Management (PIM) software, time, calculator, timer, games, and a file viewer (text, picture, and sound files were supported). In addition, optional Data Link software allowed the watch to exchange data with other programs such as Microsoft Outlook, Schedule, and Organizer. [20].

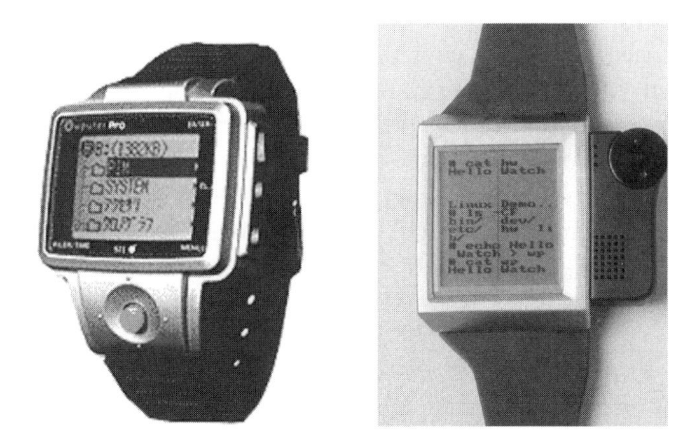

Fig. 1-4. Left: Ruputer *(Seiko Corporation of America);* Right: IBM Linux Watch ***(photo courtesy of IBM)***

In late 1998, IBM researchers in Hawthorn, New York, began to design a radically new type of watch computer [21]. Like the Ruputer, which had already been released by Seiko Instruments, the watch had a touch screen, graphical user interface, and was programmable (Figure 1-4 left). The designers also incorporated some technologies that were then usually associated with workstations. These included the Linux operating system, an Organic Light Emitting Diode (OLED) display, and Bluetooth short-range radio.

The variety of form factors in the late 1990s reflected an attempt to find the perfect physical design for the wearable computer. Although the basic form factor remained a large, fully functioning computer that typically was worn in a belly pack, we have seen that other forms were tried, including watches, and even pieces of jewelry. Most of them were either functional but not truly wearable or they were very wearable but limited and difficult to use.

One of the reasons for this difficulty was that the technology was not at a point where it could pack the functionality desired into a form factor that was truly wearable. However, there was another issue. The underlying assumption of most wearable computer designers was that you needed to replicate the power and flexibility of the desktop computer. This meant a general purpose, full functionality operating system. Most wearable computers used Linux but commercial systems also offered Windows. It also required the same suite of connectors and jacks as on a desktop system, including VGA video, serial port, parallel port, audio input and output, wired Ethernet port, PCMCIA slots, and, more recently, USB ports. As a result, most of the wearable computers in the 90s were big, bulky, and heavy. They were more 'luggable' than 'wearable'. The software sometimes also included handwriting and/or speech recognition to ease the IO chore while moving.

Toward the end of the 1990s, people started to reexamine these assumptions. Perhaps, they thought, it wasn't necessary for the wearable computer to provide all of the capabilities and peripherals of a desktop computer. Of course, this then begged the question, what should the wearable computer be? However, before answering that, they had to answer what would the wearable computer do and how would the person use it? The answers to these questions began to appear in the first couple of years of the 21st century.

1.3.2 Moving Into the Mainstream: 2000 - 2020

The early decades of the 21st century will be an exciting time for wearables. New, more wearable form factors will emerge. Wearables will become commonplace and will deliver real benefits to everyday users. There are several trends that began in the last years of the 90s and have continued into the first decade of the 21st century. These trends will enable the maturation and proliferation of wearable technology. By the early years of the second decade, wearables will become a common form of personal computing/communication. We briefly explore these trends and developments in this section.

As we have seen, wearable form factors in the 1990s were mostly confined to wearable computers or bulky watches. Today, the vast majority of devices typically referred to as wearables are still wearable computers. These are still rather bulky, although they have gotten somewhat smaller.

Recently some researchers have been adopting high end PDAs such as the iPAQ from HP. These devices typically have a StrongArm or XScale processor running at a minimum of 200 MHz, Some newer models run as high as 600 MHz and have 96 Mbytes of RAM. They usually come with a Windows operating system, although many researchers replace that with Linux. These high end PDAs offer significant processing in a small form factor. However, they typically lack many of the ports and connectors such as USB, video out, and wide area wireless transceivers that are often present in the full wearable computer.

A few years ago, Xybernaut released the Poma (POrtable Multimedia Appliance) [22]. It is a cross between a PDA and a wearable computer. It was manufactured by Hitachi and included Microsoft Windows CE operating system. It contained a processor running at 128-MHz and included 32 Mbytes of RAM. It also provided a Compact Flash™ slot and one USB port. The CF slot supported micro hard drives of up to 1 Gbyte and wireless modem and LAN cards. The USB port was reserved for a hand-held, thumb operated optical mouse pointing device included with the unit. There was no display on the unit. Instead, a monocular heads up display was included. The display provided a VGA (640 x 480) screen that appeared to the user to be the size of a 15 – 17 inch monitor at two feet in front of the user. However, at $1500, the Poma did not sell well and Xybernaut no longer actively sells it.

The leading candidate for the form factor of a wearable computer and heart of a wearable system going forward has the following characteristics:

- 400 MHz – 600 MHz 32 bit RISC processor
- 96 MB – 128 Mb flash storage
- Up to 1 GB removable memory
- 24 bit color, VGA and SVGA touch screen
- Graphics and Multimedia accelerator chips
- Audio and Video recoding and playback

Do you know what it is? Here are a couple more hints:

- You may already have one and use it every day

- It makes phone calls

Yes, for better or worse, the high end cell phone is now the leading candidate for the heart of wearable systems of the early 21st century. However, as we discuss in later chapters, it will probably not resemble the cell phone of today.

1.4 ACCEPTANCE FACTORS FOR WEARABLE SYSTEMS

Mainstream wearables are aimed at a deep integration with the user's activities and lifestyle. As such, acceptance of the technology will be a highly individualistic issue. Nevertheless, there are broad elements that are common across user populations:

1. ***Wearability***: How easy is it to put on and actually wear (as opposed to simply hang) the devices on the body; How well does it accommodate our movement as we perform our daily tasks?

2. ***Ease of Use***: How easy is it to use the devices and services, both in isolation and as part of the system; How much does it draw our attention away from what we are really trying to do when we are using it?

3. ***Compelling Design***: If the devices are visible, is it compatible with the user's sense of aesthetics? If it is invisible, is it completely unobtrusive? If the device is to be visible, it must be highly attractive. It should complement the user's sense of fashion and style, eliciting pleasure when the user or others see it. Indeed, the user may want others to see the device, either because the user is proud or excited about the device, or because it confers status – perhaps undeserved - to the user. This status may come from the design, the technology, or both.

4. ***Functionality***: Are the functions suitable for the tasks the user is performing? Is there sufficient awareness of the user's environment and situational context to enable the wearable to effectively assist the user? Do the availability and performance of the functions adequately support the user's needs?

5. ***Price***: The product must be offered at a price that reflects its value to the user. Note that for wearables, this may not reflect the value of the technology or even its function. Since wearables interact closely with the person and support their everyday tasks, other elements such as fashion, self-image, and the issue of intimacy with technology come into play.

The challenge of wearable design is to incorporate these elements into the various devices and services that make up a mainstream wearable system. We discuss each of these criterions in more detail in Chapter 5 in our discussion of wearable system design.

Whatever the form mainstream wearable systems will take, their aceptance will be determined to a large degree by the type of applications it supports. This is the focus of the next chapter.

REFERENCES

[1] Frank Moss, Director of the MIT Media Lab, Introductory remarks at the Media Lab Things That Think sponsor's meeting on May 10, 2007

[2] Shea, J. T. and Gordon, J., 2003, Wireless wearables — where's the technology headed?, *Intelligent Systems*, November 2003, http://www.sensorsmag.com/articles/1103/40/main.shtml,

[3] Venture Development Corporation, Wearable Systems: Global Market Demand Analysis, Second Edition, October, 2005.

[4] Infologix, 2006, Wearable Computers - Mobility, Portability and Handsfree Computing, http://www.infologixsys.com/products/Retail/Products/Wearable-PC/default.asp

[5] Charmed Technology – CharmBadge, 2006, http://www.charmed.com/products/charmbadge.html

[6] Xybernaut, 2005, http://www.xybernaut.com/

[7] The Cyborg Handbook; edited by Chris Hables Gray, Heidi Figueroa-Sarriera, and Steven Mentor; New York: Routledge, 1995 (pp. 29-34).

[8] Mann, Steve, 1997, Wearable computing: a first step toward personal imaging, *Computer*, IEEE. 30(2): 25-31

[9] Pentland, A. (1998) Wearable Intelligence, Scientific American, Vol. 276, No. 1, Nov. 1998.

[10] Wearable Computing at the MIT Media Lab, October 2005, http://www.media.mit.edu/wearables/

[11] The Wearable Group at Carnegie Mellon, http://www.wearablegroup.org/

[12] Family Tree of CMU Wearable Computers, http://www-2.cs.cmu.edu/People/wearable/pics/wearabletree.jpg

[13] Wearable Computing Lab, ETH Zürich, http://www.wearable.ethz.ch/

[14] International Symposium on Wearable Computers, 2006, http://www-static.cc.gatech.edu/gvu/ccg/iswc06/index.html

[15] IEEE International Conference on Pervasive Computing and Communications, 2007, http://www.percom.org/

[16] International Conference on Ubiquitous Computing, 2006, http://www.ubicomp.org/ubicomp2006/

[17] Lizzy: MIT's Wearable Computer Design 2.0.5, 1999, http://www.media.mit.edu/wearables/lizzy/lizzy/index.html

[18] Handykey Corporation, 2006, http://www.handykey.com/

[19] Strietelmeier, J., 2000, onHand PC, http://www.the-gadgeteer.com/review/onhand_pc_review, accessed January 8, 2007.

[20] Starliner's Seiko Ruputer Page, http://www.geocities.com/SiliconValley/Peaks/1559/ruputer.htm, accessed June 15, 2004

[21] Narayanaswami C. and Raghunath M.T., 2004, Application Design for a Smart Watch with a High Resolution Display, Proceedings of the Fourth IEEE International Symposium on Wearable Computers (ISWC'00), pp 7-14,

[22] Sundgot J., 2002, Xybernaut launches consumer wearable, Infosync, http://www.infosyncworld.com/news/n/1320.html, accessed January 8, 2007

Chapter 2

WEARABLE SYSTEM APPLICATIONS

2.1 CHARACTERISTICS OF A WEARABLE APPLICATION

In this chapter we discuss applications for the mainstream wearable system. Many of the applications we discuss are not the applications typically discussed for wearable systems. These applications such as virtual reality, tourist guides, and vehicle maintenance, are more relevant to specialized situations or to current wearable computers. While they are valuable applications, they do not support the activities a person would do in their typical day.

Applications that effectively exploit the unique capabilities of mainstream wearable systems will possess several characteristics not typically shared by applications created solely for the desktop.

Most obviously, the application will support personal mobility. The user wears the system on the body and thus the wearable system moves with the user and is available wherever the user is. This has several implications for how applications are designed and the type of resources they use. Among them, an application must be robust in the face of unreliable communication and unavailability of the information it may need. As the user moves, the reliability of communication with the environment and wide area data networks will vary. The current wireless data networks mostly utilize the cellular infrastructure. We are all familiar with dropped calls, lack of coverage resulting in poor quality of the communication channel with our cell phones. Wearable applications must be able to deal with this variability in communication quality and availability with minimal intrusiveness to the user.

In addition, the requirement for transparent use implies that the application recovers from the loss or unavailability of communication networks with minimal user notice and intervention. This means the application may have to suspend tasks and reschedule them for when required network resources are available, determine alternate means of acquiring the information, or proceeding without the information in a degraded accuracy or service mode. Getting the information in alternate ways could mean using other, less optimal, networks that are available.

Another possible means of dealing with unavailable network resources is to anticipate their loss and obtain required information while the networks are still available [1]. One possible method is the system keeps track of the network signal strength and as it continues to trend lower, notifies applications of this. The applications would then preemptively acquire the information needed while the networks were still available. We discuss this in more detail in Chapter 6.

The wearable application must not require the user's complete attention for prolonged periods of time. Contrast this with many desktop applications such as word processing. Since the user of a wearable system is typically doing something else while using the wearable, applications must be designed such that they convey their information quickly and clearly at a glance.

Since the wearable goes wherever the user goes, it will be subjected to much more variability in environment than a desktop PC. This includes lighting conditions, ambient noise, and mobility characteristics. No single input mechanism will be optimal for all situations the user experiences. Thus, wearable applications cannot assume nor require the use of a specific input mechanism. For example, when the user is sitting or standing still, a keyboard and GUI may be acceptable. However, when walking or driving, a speech interface may be preferable. However, if the ambient noise level becomes very high, speech is not a good option. Applications must be able to accept input from a variety of input mechanisms or, better yet, be independent of the input mechanism.

The most effective wearable applications will utilize information about the user's context. A user's context is generated by a set of data received from objects within the user's immediate environment, including sensors on the body [2]. This data is analyzed and combined into a piece of high level information relevant to the user's current situation. The union of all such pieces of information is the context. Using this context information, wearable applications can provide their services much more effectively. They can even make decisions that anticipate the user's needs. We will discuss context awareness in much more detail in Chapter 6.

Since the user is engaged in their primary task while using the wearable, excessive output by the wearable application can become annoying and be a distraction to the user. The system must render an application's output in the manner that is most effective for the current situation. For example, if the user is in a conversation with another person, the wearable may choose to queue the information until the conversation is over. Although it is ideal that an application's behavior not be dependent upon using a specific output mechanism, there may be times where it must modify its output to make it most effective given the available output mechanisms. For example, if a speech interface is used, the amount of generated speech must be carefully controlled. Users tire of hearing large amounts of synthesized speech. Also, the amount of information that can be output to the user when they are stationary is much greater than can be safely outputted when the user is in motion. Therefore, the application may be required to summarize its output to reduce it to a length that would be effective given the issues of listening to synthesized speech.

The Killer App

At this point a natural question might be: "What is the Killer App for Wearables?". You know the killer app; the application that is so compelling that it alone creates much of the market for a new technology. The term arose with the spreadsheet program VisiCalc, which played a major role in the success of the Apple II personal computer in the 1980s. The Web has often been referred to as the Internet's killer app. Indeed, the question asked about each new technology is "what's its killer app?"

Wearables would certainly use a killer app. As we have discussed, wearables have not yet caught on with mainstream users. The application of wearable technology is mostly confined to academic research and to highly vertical applications such as aircraft maintenance and courier services such as FedEx.

So wearables could definitely benefit from a killer app. Some people believe that if we just had a killer app wearables would be embraced by more vertical markets and even find their way into the mainstream. Several applications have been proposed in healthcare, home management, security, and multimedia. Surely, there is lurking in one of these the killer app.

That may not be the case. A mainstream wearable system will be very personal in its operation. Applications with the characteristics discussed above to exploit the inherent capabilities of wearables will not be applications like spreadsheets, multimedia, or web browsing. Instead they will be applications that are context aware, do not require the user's full attention, and effectively provide the right information needed for the user's current task.

Thus the real killer app may not be an application at all. It may be the total user experience of transparent, effortless access to and use of information that integrates seamlessly into the activity flow of daily tasks. Not all of these tasks are dramatic or sexy. But all of them assist us with the business of everyday life.

Nevertheless let's look at some potential compelling applications for a mainstream wearable system.

2.2 MAINSTREAM WEARABLE SYSTEM APPLICATIONS

2.2.1 Daily Activities

A mainstream wearable system will assist us in the performance of our everyday activities. The key aspect of this assistance is that the user remains focused on the task at hand, not on the operation or use of the wearable system.

Context Based Reminders

One of the potentially most useful applications for a mainstream wearable system is context based reminders. Context based reminders go beyond simple alarms/reminders based on time and date. They utilize information such as the user's location, occurrence of specific events, the current task, and environmental conditions (weather, traffic density, etc).

The rich use of context raises issues of specifying the reminders. For example, it is natural to say to another person:

- "Remind me at ten o'clock tomorrow at the airport to change my seat",
- "Remind me when I see Tom to ask him about the memo",
- "Remind Sarah at 5 pm on October first to set up the conference call"
- "Remind me tomorrow when I leave for work if it is raining to take my umbrella"

These examples each use various aspects of context beyond time and date. And each is a fairly complex command. Of course the optimal interface would be natural language speech recognition. In that case, the commands could be issued as they are given above and they are given hands and eyes free. This allows the user to specify the command in the middle of another task, greatly increasing the level of transparency of the reminder application.

However, such capability is not likely to be reliable within the near future. And as soon as we employ a GUI, we are requiring much more of the user's focus to be placed on specifying the command, significantly reducing the transparency of the reminder application.

One solution is *implicit* specification. That is, the reminder is generated from combining elements of *situational context*. For example, my wearable system could be aware of the seat assignment on an upcoming trip. By querying the airline seating database, it would learn that the seat is not an exit row. It also knows I prefer an exit row. So it creates the reminder above on its own. It also records its action, including the context elements it used, in the decision log so I can review the decision later and modify my preferences, or take other action if I consider the action the system took to be suboptimal.

As another example, from scanning my calendar my wearable knows I have an appointment down the street in half an hour. It also knows from weather reports on the web that it may rain later. It knows that I will receive a reminder in 10 minutes to leave for the appointment. Therefore, my wearable appends a reminder to take my umbrella to the reminder to leave for the appointment.

Physical Asset Management

Much of our time is spent managing our physical assets – those items we want or need with us throughout the day. Looking for one of those items when we need it can be very intrusive and interrupts the task we are trying to do.

The wearable system would keep track of these items – when we needed them, where they were and what we must do to retrieve them. Through the use of RFID tags or, if longer monitoring distances are required, short range RF technologies, our system will know what we have with us. By analyzing our calendar, location, preferences and other context elements, it will infer which ones we need and remind us if we don't have them.

For example, take something as simple as keys. If you forget your keys when you leave the house to walk to the train you take to work, you will not be able to unlock your office. Your wearable system would know that you are leaving for work since it knows you are going out the door and it knows that you go to work on the weekdays. It would determine if you have your keys with you and remind you if you did not. It could even tell you where they were and lead you to them.

Examples of research in this area is the Build Your Own Bag using RFID [3] and the Digital Paperclip (see Figure 2-1) [4] which uses a short range RF technology called 802.15.4 [5] to remind you when you are leaving something behind.

Fig. 2-1. Digital Paperclip *(Motorola Inc.)*

By utilizing context awareness, the asset management program could guard against a false positive, that is, indicating you are leaving an item behind when you either don't need it or you actually intend to leave it behind. This lack of false positives is a basic requirement for transparent use. For example, suppose you place your keys (with an attached Digital Paperclip) in your desk drawer when at work. When you leave your office to go to a meeting you do not want your wearable system to remind you that you are leaving your keys behind. You don't need them and you will be returning to your office before you leave work. By analyzing your calendar it will know that you have a meeting in the next 5 minutes (it could remind you of it) so when you leave your office it will not remind you to take your keys. However, if you are leaving the office at 5pm and you have no further meetings that day, the wearable will infer you are going home and will remind you to take your keys if you haven't already done so.

Experience Recording

Your mainstream wearable system will provide a whole new dimension to picture taking. When you take a picture it will be annotated with additional context based information that will increase the richness of the viewing experience later. This information could include the location, those with you, the environmental conditions (ambient air temperature, weather report, etc), your feelings when taking the picture, and the trip, event, or task underway when you took the picture.

When you viewed the picture, your wearable would retrieve this additional information and render it in a way that would enhance the viewing experience. For example, the wearable would display the name of the location, pictures of those with you when the picture was taken, a brief description of the trip (compiled from other sources), and an indication of the ambient environmental conditions. You would determine just how much of this information you wanted displayed and those preferences would be applied throughout the picture viewing session.

This additional information could also be used to search for photos. For example, you could request to see all of the pictures taken on a specific trip when your children were with you. Or you may want to see all of the pictures of a specific location taken at twilight.

With the rapidly increasing density of micro hard drives, it is becoming possible to record much of one's life experiences. The same algorithms that are used for annotating pictures can be applied to annotating and categorizing all kinds of media (audio, video, image, speech, sensor data, etc.) that make up our everyday experiences. To the extent that this can be done without much user intervention and the wearable can accurately determine what experiences are important enough to the user to record, this can provide a much richer set of memories of one's life [7].

Speaker Tracking

Many of us have experienced a situation in which, while we are talking to someone, we are performing a task that requires us to move away from the person. As a result, due to the distance and intervening structures (walls, etc.) between us, we can no longer carry on the conversation. Often we yell "I can't hear you" and either suspend the conversation or move back toward the person to continue speaking, interrupting the task we were performing.

With Speaker Tracking, our wearable would monitor the characteristics (distance, intervening structures, ambient noise, etc.) of the separation between us and the person with whom we are speaking. When the characteristics of the separation between the speakers match a specific connect profile, a wireless connection is instantly established between us and the person(s) to whom we are speaking without any conscious act on our part. We go from unaided to wireless communication in mid-word without missing a beat.

Similarly, when the separation between us matches a specific disconnect profile because we have moved closer together, the wireless connection is terminated, again without any conscious act on our part. We go from wireless to unaided communication in mid-word, almost transparently. This application effectively eliminates the constraints of distance on in-person communication.

Opportunistic Device Use

We are increasingly surrounded in our house, cars, and work place by media rich devices – large or medium sized TV/monitors, stereo systems and high quality speakers, etc. At the same time our cell phones and PDAs are handling increasingly richer media content – videos, high resolution images, and stereo music. The experience of listening and/or viewing this media rich content is often compromised by the limitations in the audio and video capabilities of our cell phones and PDAs (small screens, small speakers, etc.).

The mainstream wearable system will seek out and utilize the devices in our immediate environment that are capable of optimally rendering media rich content. For example, if we have a high resolution video that we want to view and we are near a large video screen in our home, the wearable system will realize this and send the video to the large screen which will provide us with a much better viewing experience than if we viewed it on the small screen of our wearable.

The crucial aspect of this will be its transparency. The wearable system will become aware of the presence of a device that can better display the video we are watching on the small screen of our wearable, determine that it is idle, and initiate a connection with it all without our intervention. It begins streaming the video to the monitor, reformatting the video to best fit on the monitor. If we were to walk away from the monitor, the wearable system would automatically revert to showing the video on the smaller screen of our wearable device, again, without our intervention.

Hands Free, Eyes Free Web Surfing

One of the most important services a wearable system will provide is the transparent access to and utilization of information to aid us in the current task. The Internet is the most extensive information source ever constructed. However, browsing it usually entails sitting in front of a computer and traversing links from one page to another. It is currently a highly visual and attention focusing task.

However, with a mainstream wearable system you will be able to surf the web and obtain information almost transparently. You will speak a topic description into your earpiece and the wearable will search the net for the most relevant pages. It will use information about your interests, current task, context, and preferences to narrow the search to the most relevant pages.

When your wearable receives a web page it extracts the text, ignoring all of the visually oriented material. It then renders the text using speech synthesis in a male voice, summarizing it if the user's context requires it. When it comes to a hyperlink in the text, it reads the link text in a female voice. This change in voice in the speech synthesis signals the user that the text is a hyperlink. The user can speak any substring of the hyperlink text and the wearable will retrieve the web page associated with that link. This enables the user to surf the web solely by speech [6].

Opportunistic Communication

Opportunistic communication is defined as communication that is initiated, only because it is trivial to do so. Once wearable systems become truly transparent to setup and use, people will utilize short range networks such as Bluetooth to engage in communication among people separated by short distances.

For example, if you wanted a specific website and you knew some of the URL but not all of it, you probably would not pull out your cell phone and call someone to ask what the URL is. Instead, you might try to search for it yourself. The effort required to make the call is not worth it if you think you are likely going to be able to find it. However, after searching in vain for ten minutes, you give up and make the call.

Now imagine if you could simply say "John, what is the site URL?" Your wearable would recognize the name 'John' as the name of a person and look in the alias phone directory to select the phone number of the person that is identified as 'John'. It buffers the rest of your utterance ('what is the site URL') and connects. When John answers, the wearable plays the entire utterance. John now answers the question and you have your information. You say thanks and your wearable terminates the call.

Notice that you did not have to handle the phone, nor remember any phone numbers. In fact, there was no discernable action on your part (except for your utterance) that involved remote communication. You 'made the call' as the first recourse to finding the URL only because it was trivial to do so. This is the essence of Opportunistic Communication.

2.2.2 Cognitive Assistance

All people suffer temporary, situational cognitive impairments. For example, heavy multitasking will often result in forgetting a task or appointment. A mainstream wearable system can assist its user by providing information that directly addresses the current task, either proactively or as a result of a user's request. By monitoring the user's activities and interactions with the wearable, it may be able to determine when specific help is required.

Acquaintance and Situation Recall

Most of us have experienced the awkward situation where we meet someone we have met before. They remember us, but we cannot remember them or their name. And so as the conversation with them progresses, we try to elicit clues to their name while trying to hide the fact that we don't actually remember it.

The wearable system would acquire information about the person approaching you. This could be via a small wearable camera and face recognition or receiving an infrared or short range RF transmission from the person with information, including their name, they wish to make public.

Your wearable system would check your acquaintance database for a match and, if found, would send the person's information to you via some private mechanism such as an wireless earpiece or a display in your glasses.

To prevent notifying you of people that are well known to you, the wearable would maintain information about the last time you met this person, their relationship to you, and other information used to estimate how familiar you may be with the person. The system would not remind you about the identity of those people whom you are likely to be familiar.

However, to recover from false negatives, or if there is no information about the person in the acquaintance database, the system would open its far field microphone and, using speech recognition with keyword spotting, listen to see if you requested the person's name. If you did ask for the person's name or the person offered it unsolicited, the system would take a picture of the person and create a record in the acquaintance database. There are obvious privacy issues that must be addressed and these are discussed in Chapter 7.

Once the person's name is provided to you the system could also provide information about the last time you met, the event and location of the last meeting, and some of the topics discussed. This information could be acquired using the wearable's speech recognizer with keyword spotting during the interaction with the person[4].

This application can also be used to familiarize you with the people you are likely to meet at an event before you even go. If you know the names of the people who may be there but can't remember their faces or anything about them you can manually query the acquaintance database for the person's picture and information. Then, should you meet them at the event, you would have no problem remembering them and engaging in a conversation.

Of course, this application could also be used to help you avoid people with whom you did not want to speak or interact. By retrieving their face and information you will be reminded why you do not want to speak or interact with them. Then, at the event, you are better able to avoid them.

Entries in the acquaintance database would undergo aging and those entries that have not been accessed for a specific period of time (which the user specifies) would be deleted. Also entries corresponding to events that do not fit your interest profile would also be deleted. This reduces the number of records corresponding to people that you meet only once.

2.2.3 Task Management and Planning

By managing the many repetitive and straightforward tasks a person faces in their daily life, the wearable system will allow its user to concentrate on those activities and tasks that are most important to them. In addition, using

[4] An early version of this is the Remembrance Agent. The Remembrance Agent (RA) is a program which augments human memory by displaying a list of documents which might be relevant to the user's current context. It runs continuously without user intervention. When the user encounters a situation or person he remembers later, he can manually enter notes about it. Later, when they encounter it again, the application would display the information previously entered.[7]

technologies such as goal planning and game strategy, the wearable would be able to plan optimal approaches for the user to employ in completing their tasks. Examples are appointment scheduling, path planning for shopping, and sequential task completion.

A wearable system can assist the driver by monitoring their vital signs for fatigue, stress, and fear. Once the wearable knows the driver is in the car, it can use the vital sign information to recommend actions to the driver. For example, if the wearable detects the driver is tired, it can recommend that he reduce speed or pull over at the next rest stop. The wearable could use its networking capability to find the nearest hotel and direct the driver to it by the route most appropriate for the driver's current condition. If the wearable can also interface into the car's sensors, as would be the case in emerging telematics systems, the wearable can send instructions to the car. For example, the wearable could instruct the car to raise the volume of the car's audio system to help keep the user awake.

Effective but Humane Marketing

The scene in the movie 'Minority Report', in which John Anderton (Tom Cruise) is walking through a mall and is being bombarded with unsolicited ads of nearby stores, is enough to make anyone oppose presence based eCommerce. However, properly and sensitively done, such context based marketing can be much less intrusive and very useful.

A mainstream wearable system would mediate the onslaught of ads sent to the user from nearby stores. The system would determine what store sent the ad and whether the store sells items that meet the user's interest profile. The wearable system might even be able to inspect the contents of the ad and determine if the actual product being advertised meets the interest profile.

The interest profile would be context sensitive. It would take into account your current task, your current location, the set of tasks to be done at the location and also within a specific time (say the next four hours). It would maintain a list of gifts or other items you must buy based on the contents of you calendar and emails. It would also keep track of when you last purchased a product from this store and how frequently you shop there.

All of this information would be used to evaluate how relevant and important the ad is to you. Ads that pass review would be forwarded to you using the output medium most appropriate to your current situation. This could be audio, a message on your GUI device, or a message overlaid on your display glasses.

As with all decisions the wearable system makes, the decision process for the ad would be logged, allowing the user to inspect it upon demand and determine if the decisions made were proper and optimal. If not, the user can alter the criteria used by the interest profile and ad review application to better align it with the user's desires.

2.2.4 Health maintenance and support

Health and wellness maintenance will be one of the most promising areas for wearable applications. The improvement in small, low power sensors and in data fusion for context awareness will allow a wearable system to monitor several body vital signs and detect anomalous readings. The system can relay these readings to a central support agency and/or interact with the user to give advice.

Personal Coach

Many people find it hard to maintain the motivation for continued, regular exercise. The user's mainstream wearable system can increase their chances of maintaining an exercise program by providing context based feedback and exercise status.

Sensors worn on the body or embedded in workout clothing will monitor the user's heart rate, blood pressure, pulse, and body temperature to provide a snapshot of the user's exertion level. Other contextual information such as location (gym, health club, etc) time of day, and even which machine you are on or your speed of locomotion in the case of running or biking provide information on the specific activity the user is engaged in.

All of this information will be analyzed in the context of the user's workout plan. The application will keep track of exercise duration, user performance (machine level setting, run time, etc) and track the user's progress over time. This progress could be compared to the goals the user has set or a doctor has recommended.

The goal vs. progress would be tracked and presented to the user upon demand or after each workout to further motivate the user. In addition, anomalous or abnormal events (severe drop-off in exercise intensity and/or duration) would be flagged for user review and possible doctor notification.

The personal coach application would also help maintain the user's motivation while exercising. If the user is listening to music while exercising using the sensor based and context based data described above, the application could adjust the music tempo, or even select a new song to better match the song's characteristics (tempo, volume, etc) to the planned exercise intensity and/or duration. For example, if the user is starting to run slower and the exercise plan calls for maintaining the faster pace, the current song's tempo could be increased or a new song with a faster tempo selected to help the user maintain the planned running/biking speed.

During exercise dehydration is a common, and potentially serious, condition. If the user is exercising outdoors during a hot day dehydration can contribute to heat stroke, which can be fatal. The personal coach application can monitor the user's core body temperature, the ambient air temperature, skin conductivity (correlated to amount of sweating), and remind the user to drink water if the user is becoming dehydrated. It could also warn the user to slow down or even pause if the sensor information indicates conditions are ripe for the occurrence of heat stroke.

Mood Manager

Many of us do things in the heat of the moment that we later regret. If we are lucky, the consequences of the rash action are, if not inconsequential, at least transitory. The old saying of 'count to ten before you act' has real merit.

A wearable system can help motivate that period of reflection before action. By detecting rising levels of frustration and understanding the context in which they occur, the wearable can take action targeted at reducing the frustration level and giving the user a better chance to think before acting.

Sensors worn on the body or in clothing would register rising blood pressure, faster heart rate, and increased skin conductivity. Sensors could also record increased muscle tension in the arms. Speech recognition could recognize obscenities and the tone of the voice analyzed for signs of anger (loud, forceful speech).

In some cases, the user's environment could be instrumented to help detect rising anger and frustration. Steering wheels in cars could contain pressure sensitive material to detect when the user is gripping the wheel very tightly. Other areas of the car (dashboard, center seat consoles, etc) can contain impact sensors to detect when the user strikes them in frustration. This data would be sent via Bluetooth or other appropriate short range wireless protocol to the wearable for analysis.

Once the wearable detects anger and/or frustration, it can take action to reduce it or take other actions to compensate for the user's condition. For example, the wearable can start playing music the user finds soothing. Or it can instruct the car the user is driving to slow down to compensate for the driver's likely reduced concentration on the road.

2.2.5 Personal Security

Personal security is becoming more important every day. While cell phones can enable us to call for help wherever we are, a wearable system can help prevent us from getting into danger or provide information to help reverse or minimize the impact of a dangerous situation.

Personal Radar and Witness

The personal witness application utilizes contextual information to anticipate danger and warn us. The wearable will know where we are, the amount of ambient light, time of day, and who we are with. If it is dark and we are walking alone, the wearable will retrieve recent crime statistics for the area. If they indicate recent crime activity, the wearable will go to 'high alert'.

Under high alert, the wearable would send periodic notifications of your location and estimated time of arrival (ETA) at an intermediate point in your trip to a 'trusted person' such as a parent, spouse, or friend. These intermediate points would be determined by using a route planning algorithm and Geographical Information System (GIS) database[5]. The wearable would calculate the ETA at the intermediate point using your traveling speed as determined from the change in GPS coordinates averaged over a period of time. The wearable would also send a notification when it reached the intermediate point. If the trusted person's wearable system did not receive the notification that you arrived at the intermediate point some period of time after the ETA, it would notify the trusted person who could then call E911 on your behalf or take some other action to help you.[6]

A small, low power thermal imaging system could form an important part of a wearable system's personal security system. These systems can detect a person's presence based on the difference between their body temperature and the surrounding air temperature using infrared cameras and image processing software.

When the wearable goes on high alert, the imaging system is activated. It can detect the presence of a person near you and perhaps even indicate if they are moving toward or away from you. Based on their movement and location, your wearable could warn you to take some action to avoid coming too close to this person or to send out a call for help.

[5] A GIS service converts GPS data (latitude, longitude, and altitude) into the nearest physical building, landmarks, or other points of interest. For an example, see [9]

[6] The functionality described in this application has been prototyped as part of the 'Will You Help Me' project at the MIT Media Lab [10].

However, even with the most sophisticated equipment, it may be impossible to avoid being the victim of a holdup or other crime. However, your wearable system can help here too. The use of deform-sensing material woven into the clothing would detect the position of the user's arms. Sensors on the body would track the change in the user's pulse and blood pressure.

The wearable system maintains a series of distress profiles. These profiles would contain specific data on the position of the user's arms and sensor values. When the actual readings of the sensors matched a specific distress profile, the wearable initiates a silent E911 call. Then the wearable would activate a hidden far field microphone and a concealed video camera. The wearable would contact a monitoring company and begin streaming the audio and video to the company that stores it for future use, perhaps in a trial when the perpetrator is caught.

The mere knowledge that a person may have such a system could provide a powerful deterrent to would be attackers, further enhancing the user's personal security.

2.3 MAKING THESE APPLICATIONS TRANSPARENT

Most of these applications utilize a significant amount of contextual information (location, blood pressure, muscle tension, ambient temperature, etc.). There are several issues with acquiring and processing this information that must be resolved if these applications are to be transparent as required for a mainstream wearable system. These include

- Acquiring incoming context data and processing it into a piece of high level, relevant information;
- Presenting the information to the user in the most effective manner as dictated by the system's awareness of the user's current situation;
- Degrading gracefully when the required data is not present or complete for the information needed by the user.

We discuss these issues in the following chapters.

REFERENCES

[1] Balan, R. K., 2004, Powerful change part 2: reducing the power demands of mobile devices, PERVASIVE computing.3(2): 71–73.

[2] Filho R., 2002 Awareness and privacy in mobile wearable computers. IPADS: interpersonal awareness devices. Final report, ICS-Information and Computer Science, 319 UCI--- University of California, Irvine. http://awareness.ics.uci.edu/~rsilvafi/papers/VirtualColocationFinalPaper.pdf , accessed March 20, 2006.

[3] Nanda, G., Bove V. M. Jr., and Cable A., 2004, bYOB (Build Your Own Bag): A computationally-enhanced modular textile system, http://alumni.media.mit.edu/~nanda/design/electronics/byob/papers/bYOB_UbicompDemos.pdf, accessed March 20, 2006.

[4] Ferenczi, P. M., 2006, Super Prototypes, Laptop Magazine. 24(8): 99.

[5] IEEE Standards, 802.15.4T, 2003, IEEE Std 802.15.4™-2003, IEEE Computer Society, May 12 2003, available at http://standards.ieee.org/getieee802/download/802.15.4-2003.pdf, accessed March 20, 2006.

[6] Audio interface for document based information resource navigation and method therefore, US Patent 5,884,266, March 16, 1999.

[7] Maney K., 2004, Every move you make could be stored on a PLR, USA Today.com, http://www.usatoday.com/tech/columnist/kevinmaney/2004-09-07-plr_x.htm

[8] Rhodes, B.J., 1997, The wearable remembrance agent, Proceedings of 1st International Symposium on Wearable Computers, Cambridge MA, IEEE Press, 123-128.

[9] Converting Latitude/Longitude Coordinates into GIS Points, Natural Resources Conservation Services, US. Department of Agriculture, http://www.nrcs.usda.gov/intranet/rad/point.html

[10] Chung, J., 2006, Will You Help Me, http://www.media.mit.edu/speech/projects/WillYouHelpMe.html

Chapter 3

OVERVIEW OF WEARABLE SYSTEMS

3.1 WHAT IS A MAINSTREAM WEARABLE?

In the previous chapters, we discussed the history of wearable systems and some potential applications for a mainstream wearable system. But what exactly is a mainstream wearable system? More fundamentally, what is a wearable system? What characteristics make a system wearable? Simply strapping a laptop onto your belt is hardly an acceptable implementation. We need to define the salient attributes a computing/communication system must have before we call it a wearable system.

There are many possible definitions of a mainstream wearable system. Most fundamentally, we can ask, is any device that is attached to the body a wearable? That is, does the simple fact of wearing something make it a wearable?

Figure 3-1 shows three devices that can be worn on the body. Are they 'wearables'? Are they mainstream wearables?

We can also ask, is anything embedded or contained inside clothing a wearable? Embedding electronics into garments has been researched for some time now and there are several approaches that have been tried.

Figure 3-2 shows three different levels of embedding devices into clothing. The Scott eVest [1] has as many as 40 pockets into which you may place devices. It also has internal wiring guides to route earphones and microphones to your cell phone or mp3 player. Is this vest a wearable?

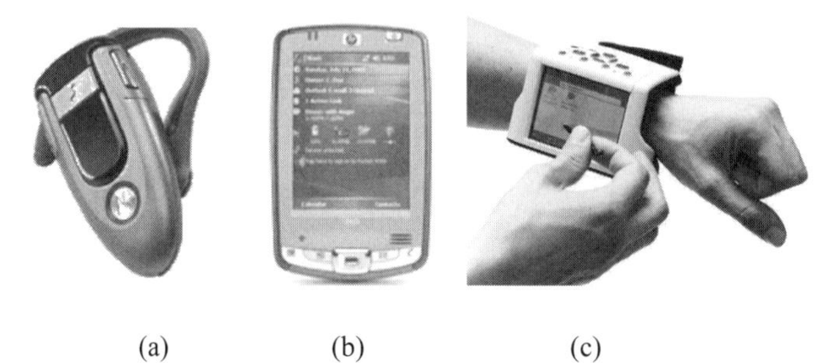

(a) (b) (c)

Fig. 3-1. Potential Wearable Devices: (a) Motorola H500 Bluetooth Headset *(Motorola Inc.)*, (b) HP iPAQ PDA, (c) Zypad WL 1000 *(EuroTecH S.p.a., , www.eurotech.com and www.zypad.com)*

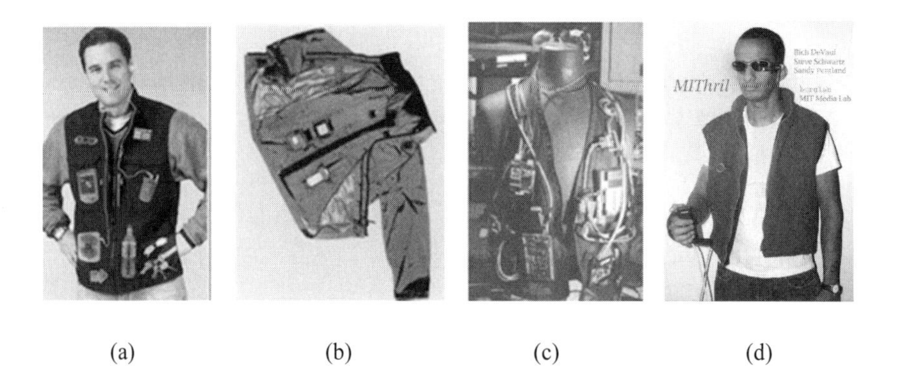

(a) (b) (c) (d)

Fig. 3-2. Devices Contained or Embedded in Garments: (a) Scott eVest (Scott eVest, *www.scottevest.com*), (b) ICD+ jacket (*All rights reserved, Philips Electronics*), (c) MIThril vest inner view(MIT Media Lab / Alex Pentland) , (d) MIThril vest with overgarment (MIT Media Lab / Alex Pentland)

The ICD+ jacket from Levis [2] also has compartments for devices and wiring for earphones and a mic[7]. However, unlike the eVest, the devices are

[7] The jacket is no longer available and its website has been taken down.

not simply placed into pockets. There is a sophisticated connection mechanism and the devices are embedded within the jacket with internal wiring for control of the mp3 player and cell phone. Is this a wearable?

The Mithrill vest from the MIT Media Lab [3] shows the deepest level of embedding of the three. Here, individual components (processors, internet browser engine, Ethernet controller, WiFi transceiver etc.) are distributed throughout the vest. Is this vest a wearable? The rightmost picture in Figure 3-2 shows the vest inserted into its fabric over garment. Is the vest now a wearable?

Is a wearable computer really a wearable? At first this seems like a foolish question since its very name should settle the issue. Indeed, as we saw in the history of wearable computers in the first chapter, this is where the most common notion of a wearable comes from.

Figure 3-3 shows two wearable computers. Of the two, only the POMA, the computer on the right, was aimed at consumers. And while the other one is a full fledged desktop equivalent, the POMA [4] is more akin to a PDA whose emphasis is on internet access and web browsing.

Are these wearable computers really wearable? The answer is probably yes since many of the devices making up these systems conform to the shape of the part of the body on which they are worn. However, are they mainstream wearable systems?

Finally, is any piece of jewelry with embedded electronics a wearable? They are certainly worn on the body. Figure 3-4 shows two pieces of what would be commonly labeled as jewelry, each with embedded electronics. The left picture is the Nokia Medallion I [5]. It allows you to upload and display pictures to it from your phone via an IR connection.

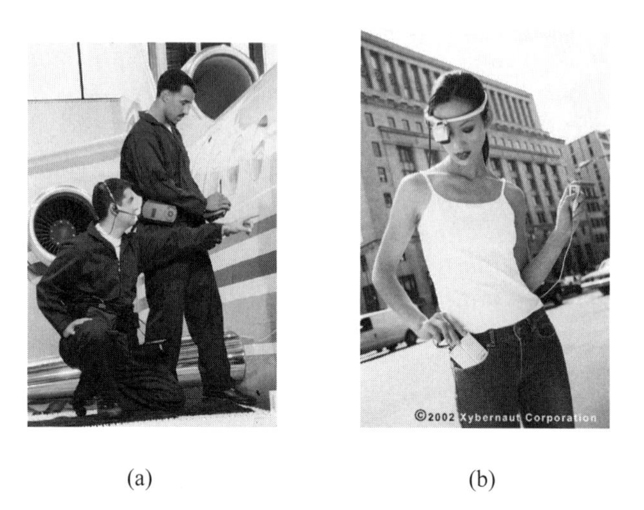

(a) (b)

Fig. 3-3. Wearable Computers: (left) Xybernaut MA V *(Xybernaut Corp.)*, (right) POMA *(Xybernaut Corp.)*

The right picture is a ring containing the iButton from Dallas Semiconductor [6]. It contains a small sensor that sends its information when interrogated by a reader. It can be used for access control.

Not shown is a $500,000 gold/platinum Heartthrob Brooch with a series of LEDs [7]. Behind the broach is a circuit board with a heart monitor and low power radio transmitter. The broach detects the user's heart rate and sends it to a PC. The LEDs blink to the pace of the heart. Are these pieces of jewelry wearables? Are they mainstream wearables?

The problem is that the current notions of a wearable are very imprecise. In fact, we can make several observations about the current concept of a wearable from the pictures we have seen:

- Wearable does not imply small. Most of the wearable computers and examples of clothing are quite large.

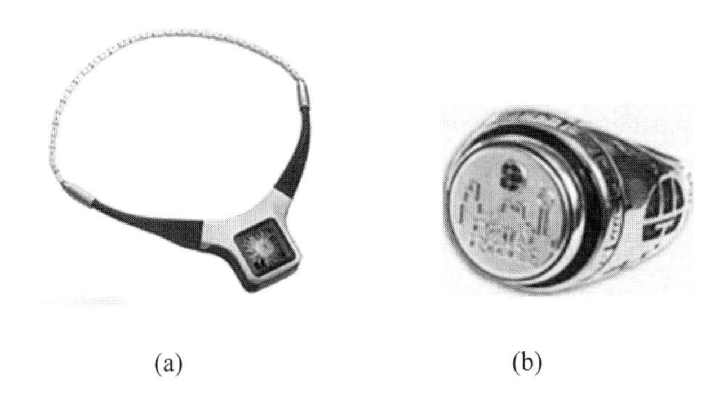

(a) (b)

Fig. 3-4. Jewelry with Embedded Electronics: (a) Nokia Medallion ***(courtesy Nokia Corp)***, (b) iButton® ***(photo courtesy of Dallas Semiconductor)***

- Wearable does not imply easy to set up. Many of the systems shown above involve several components and connections required to set up the system for use. Even jewelry can be difficult to put on, to which anyone who has fumbled with the very small, intricate clasps and hooks of fine jewelry can attest.
- Wearable does not imply easy to use. Most of the wearable computers in Figure 3-3 do not look easy to use and in many cases they are not. The ICD+ jacket in the picture second from the left in Figure 3-2 has several pages of instructions detailing how to remove the embedded devices before cleaning the coat.
- Finally, wearable does not imply attractive. Most wearable computers and even other wearable devices are very utilitarian looking. This is mainly due to the fact that they are aimed at vertical applications in industry and functionality is paramount. Attractiveness is not a main consideration.

Of course, there are many counterexamples for each of these observations. However, the point is that the term wearable as it is currently used is ambiguous. If we are going to discuss mainstream wearable systems, we need to have a more precise definition of what a wearable is.

3.2 CHARACTERISTICS OF A WEARABLE SYSTEM

In 1998, Steve Mann specified the characteristics of a wearable computer [8]. Many of these are applicable to mainstream wearable systems, either directly or with modification.

According to Mann, a wearable computer exhibits the following characteristics:

UNRESTRICTIVE: Wearable computers do not restrict your mobility. For example, you can interact with the wearable computer while walking, sitting, navigating constricted spaces, etc.

UNMONOPOLIZING of the user's attention: The display and user interfaces of the wearable computer do not require your complete attention. You can perform other tasks while using it. Unlike desktop systems, a wearable system is used in the background, allowing you to concentrate on what you really want to do.

OBSERVABLE by the user: As a rule, a wearable computer will be operating the entire time you are wearing it. You can view its displays, and input commands whenever and wherever you are. In addition, the computer can get your attention whenever necessary.

CONTROLLABLE by the user and responsive: Despite the fact that the wearable computer is always operating while you are wearing it and performs many of its functions in the background, you can intervene and take control of it at any time. Any operation the computer is doing can be aborted or modified by the user.

ATTENTIVE to the environment: The wearable computer has the ability to communicate with and interact with elements in your environment. This includes exchanging information with devices such as PCs, printers, etc. It also includes sensing the characteristics of you and your environment. Examples include monitoring your blood pressure, heart rate, etc and sensing the temperature of the air around you.

COMMUNICATIVE to others: Almost every wearable computer would include the ability for wide area communication, whether using the cellular infrastructure or leveraging off of available high speed wireless local area networks that provide access points to the cellular or wireline network.

CONSTANT: Always ready. The wearable computer is always capable of interacting with the user. Note that this does not preclude low power sleep modes for power management. However, the computer should transition from its sleep mode to active mode quickly.

PERSONAL: The computer is typically the exclusive property of its user. The user and computer interact in a closely directed manner. This implies the following capabilities:

- PROSTHETIC: The computer can act as a true extension of mind and body. The mode of interaction is natural and after time you forget that you are wearing it.

- ASSERTIVE: The wearable is considered part of you and as such may be harder for others to request that it be removed. Contrast this with being requested to leave your laptop or camera at the entrance to a store or place of business.

- PRIVATE: Others cannot interact with the wearable computer unless you permit them. In addition, it may be difficult for another person to tell if you are using it. As an example, you may be able to capture video or images with a wearable camera as part of the wearable computer or communicate with a remote party without a person near you realizing it.

These characteristics are very device oriented. Our definition of a mainstream wearable system, while incorporating (and sometimes modifying) the above characteristics, is more user oriented:

A wearable system is a collection of devices worn on a person's body that seamlessly, and always under the control of the user, collaborate to assist the user in everyday tasks. These devices, both separately and together, have little or no Operational Inertia and are proactive and non-intrusive in their operation. The user employs these devices in an almost unconscious manner, realizing an increase in the quality of life.

The requirement of zero or near zero Operational Inertia (OI)[8] means that:

- There is little or no setup effort required to get the devices ready for use,
- The use of the devices is intuitive to the user, and
- The user is rarely, if ever, aware of the presence of the devices as she goes about her daily tasks.

This definition does not mention many of the characteristics of Mann's definition. However, the benefits of those characteristics are present in the OI requirement. Take, for example, Mann's requirement that the device(s) be always ready. The benefit is that they are always and instantly available to the user. This is implied by the low setup effort of the OI requirement.

Similarly, the devices cannot restrict the mobility of the user because they must be completely, or almost so, unobtrusive to the user so that the user is never, or rarely, aware of their presence.

Ours is a rather stringent definition and would disqualify many devices now being described as wearable shown above. It clearly disqualifies most of the systems that today's cyborgs wear. These systems usually consist of

- A computer worn on the belt or in a bag draped around the user,
- Some type of heads up display, usually a small LED screen just in front of one of the user's eyes like a Private Eye, and
- Some type of input device, typically a one handed chording keyboard like a Twiddler [9], shown in Figure 3-5.

These systems can be very obtrusive, often intrusive, and could not in any way be considered part of a person's normal apparel.

[8] Operational Inertia is formally defined in Chapter 4.

Fig. 3-5. The Twiddler *(Handykey, Inc.)*

This definition also disqualifies most of today's communications devices. Consider many of the new phones. They are very small and, most of the time, quite non-obtrusive. However, their small size can make them intrusive as the user must concentrate on which small button is pushed and must still hold the phone to the ear to use it.[9]

The important thing to realize about a mainstream wearable system is that it is not a wearable computer. Its focus is on seamless integration into a person's daily task flow, being unobtrusive and not constraining the user's movements.

It is also not a replacement for the desktop PC. Its focus is task and user augmentation. Most applications will be used in support of another task that is primary to the user. Nor is it aimed at just industry, public safety or the

[9] A cell phone with voice dialing and a Bluetooth headset approaches the characteristics of a mainstream wearable system, at least for making a phone call. Other factors such as the non-use obtrusiveness of the Bluetooth headset and the interaction complexity of the voice dialing (in large part determined by the accuracy of the speech recognizer and ease of correction of misrecognitions) determine if it actually reaches mainstream wearable status.

military. It is applicable to all population and market segments if properly designed.

So our challenge is to find ways to create devices that provide computing and communications service like many of the devices today, but in a way that allows them to be almost transparent in their use.

Before we go off designing, let's look at the different ways a mainstream wearable system interacts with the user and vise versa. Understanding this will give us insight into how to best design the system's user interfaces, services, and device form factors.

3.3 USER – WEARABLE INTERACTION MODES

To understand the interaction between a mainstream wearable system and the user, it is important to view the user and the wearable as a single, larger system. The interactions between the user and the wearable devices take place within this system. Thus, we should not view the user and the wearable as two separate entities that interact independently. Rather, there is an element of collaboration and even dependency between them. The possible inputs and outputs between the user and the wearable system are shown in Figure 3-6 [10].

Input from the environment external to this system comes both to you and to your wearable[10]. This means that the wearable competes with the external environment for your attention. This has implications on how the wearable delivers information to you and how it gets your attention to indicate that you have information waiting or that it requires your input.

[10] Unless otherwise noted, when we use the term 'wearable', we include all of the devices in the wearable system taken as a whole.

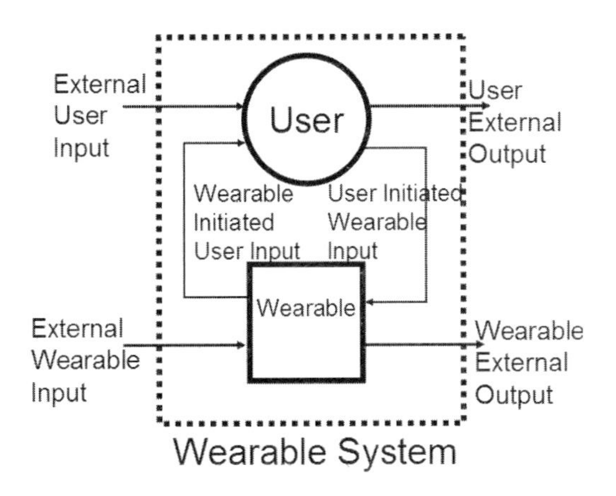

Fig. 3-6. User and Wearable Computer as a System (adapted from [10])

One area most affected by this is the design of the GUI for a wearable's display, if it has one, and the design of a speech synthesis system if it is part of the Speech User Interface (SUI). One challenge is ensuring that when the wearable conveys information to you, you will pay attention to it. This may require that you reduce your focus on what you are doing to attend to what the wearable is going to tell you. One possible mechanism of getting the user's attention is to precede whatever information the user will convey with an audible alert. These can be simple tones, short musical sequences, etc. Another approach is to precede the information with the user's name. This is particularly applicable when the wearable uses speech synthesis as the output medium. So instead of saying, "You have a meeting in 5 minutes", it says, "Joe, you have a meeting in 5 minutes. This is effective because it is known that people respond to their own names and will change focus to whatever mentioned their name [11]. However, this has to be used judiciously since people are very good at filtering out repeated stimuli.

Similarly, you must compete with the external environment for the wearable's attention. This influences the design of the input mechanisms used by the device including the speech recognition system. If the environment is noisy, or the computer is busy processing sensor data from the environment, there must be a way for you to reliably and quickly get the wearable's attention if you want to give it a command or reply to a query.

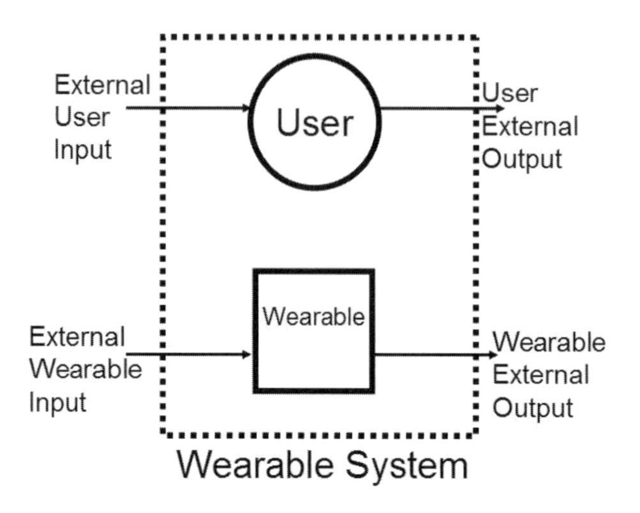

Fig. 3-7. User and Wearable Acting Separately

This means that there must be multiple input mechanisms since one interface mechanism will not suffice for all situations.

It is this competition for attention by both the user and the wearable for input and output that poses some of the biggest challenges in wearable system design. Let's look more closely at the ways the interactions between the user and the wearable can occur.

There are times when there is no collaboration between you and your wearable (Figure 3-7). When acting alone, you receive input directly from the environment and respond directly to it. The wearable is not involved with what you are doing, although it may be operational and working on its own tasks that do not require interaction with you. In such situations, it is desirable that you be aware of the wearable as little as possible since there is no need for interaction and any awareness would likely distract you from your task at hand.

Your input and output can be mediated by the wearable. There are many ways your input can be augmented by the wearable. In the simplest case, shown in Figure 3-8, the wearable overlays information onto what you see. This is the case with augmented reality applications. The information from

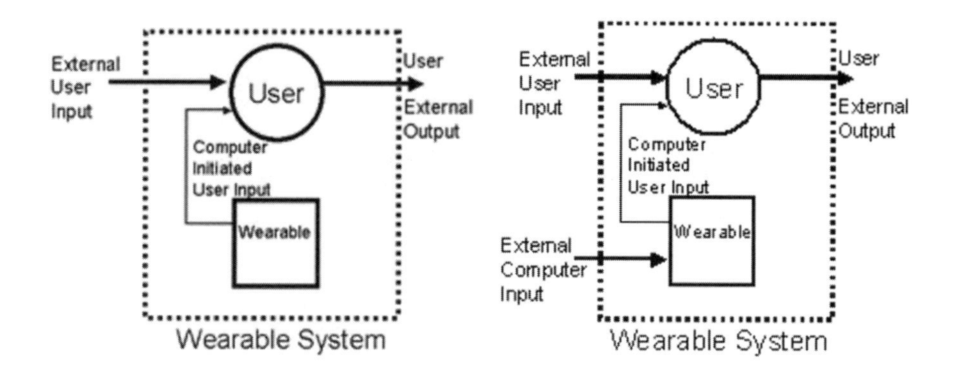

Fig. 3-8. User Input Mediated by the Wearable System (adapted from Mann, 1998)

the wearable can be local to the wearable, as in the case of a tour guide application where all of the information about the city is on the wearable's hard drive, shown on the left of Figure 3-8. Alternatively, the information overlaid can come to the wearable from the outside. An example would be a battlefield tactical display, where the user's view is overlaid with information received by the wearable over a secure wireless link. This information could include the locations of members of the user's squad, or information about enemy forces received from aerial observation or other means. This is shown on the right of Figure 3-8. Many augmented reality systems will contain a combination of sources for the wearable's overlaid information.

An interesting variation on augmented reality is the EyeTap system designed by Steve Mann **Error! Reference source not found.**, [13]. EyeTap is a unique hardware and software combination that can be anything from unaided vision to immersive Virtual Reality.

When wearing the EyeTap hardware shown in Figure 3-9, the eye is completely occluded by the device. EyeTap contains both a camera and

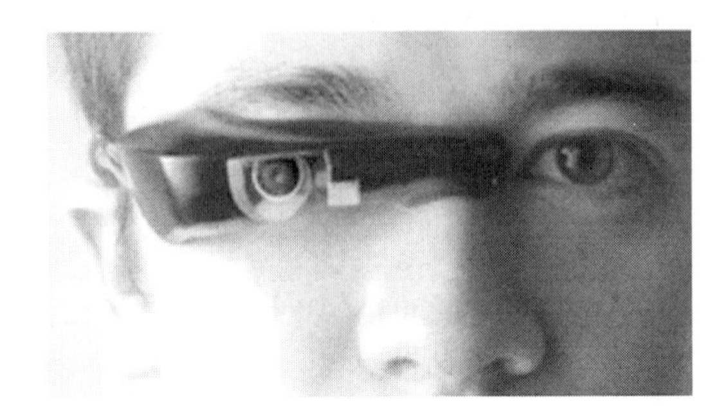

Fig. 3-9. The EyeTap Device *(Copyright Steve Mann, 1998)*

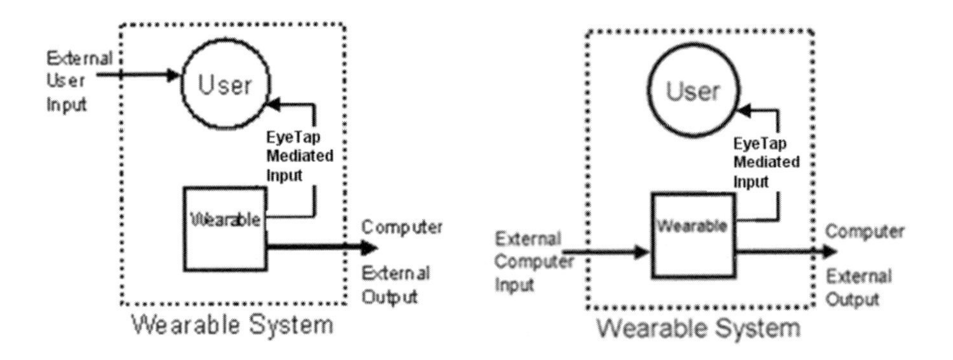

Fig. 3-10. User Input Mediation by EyeTap

display that work closely together. All input to the eye is intercepted by the camera and sent to EyeTap's Mediated Reality system where it can be processed in any way the user desires. This processed (or unprocessed) image is then sent to the eye via the Aremac[11]. Thus, EyeTap interposes

[11] 'Aremac' is' camera' spelled backwards.

itself between the real world and the user. This is shown on the left in Figure 3-10. If the eye without EyeTap is closed, then there is no direct visual input to the user as shown on the right of Figure 3-10. All of the visual input is received by the wearable via EyeTap, processed, and then sent to the user. The only visual user output is that provided by the Aremac, part of the EyeTap hardware[12].

3.4 FORM FACTOR OVERVIEW

When we talk about wearables, most people think of wearable computers. Wearable computers have a distinctive form factor. Basically, they are desktop computers scaled down as small as possible. The amount of size reduction is limited by the philosophy most wearable computers embody. This philosophy is to provide all of the capability and options typically found on a desktop computer. As a result, they must include the physical ports currently found on most PCs, including video out, serial port, 2 or more USB ports, audio ports, Ethernet port, Firewire (IEEE 1392) port, and at least one PCMCIA card slot. These physical ports take up space and limit how small the wearable computer can become.

3.4.1 The Rush to Integrate

Another issue with wearable system design is the desire to create a single, integrated device. This is reflected in the trend toward convergence – packing more and more functionality into a single device. Cell phones are perhaps the clearest example of this trend.

[12] For the full potential of mediated reality, the user must wear an Eye Tap on both eyes. Otherwise the user receives input directly from the environment via the uncovered eye. This input cannot be processed by Eye Tap's mediated reality processing.

However, is the integrated form factor the best path to pursue? Integrated devices are most successful when:

1. There is a clear consensus on the suite of devices and functionality that people wish to have in a single device
2. The integrated functionality can leverage off of one another and this leverage is difficult or impossible to achieve without physical integration
3. The physical integration does not significantly increase the complexity of operating the device
4. The technologies underlying the integrated functionality are stable and slowly evolving so that the user feels his investment is protected

Do these conditions hold for mainstream wearable systems? Consider:

- There is no clear consensus on the suite of devices/services that most people want integrated into a single device. Indeed, the rapid introduction of integrated devices with different suites of services and device capabilities resembles the Cenozoic Era in Earth's prehistoric past in which Nature experimented with numerous variations in life forms in an attempt to see what designs and capabilities worked best. Most of those creatures failed to survive and became extinct. It is clear that, for some combinations of devices, there is significant leverage to be gained by integration. Integration of a digital camera or a GPS receiver with a mobile phone is an example. But beyond that, it is not clear which of a large number of potential suites of devices provide significant user appeal in an integrated device. For instance, here is a list of some of the functionality put into phones at one time or another, either as a product or prototype:

Camera	GPS
Fingerprint Reader	Barcode Scanner
Breathalyzer	Tilt and Motion sensors
TV tuner	Projection display
Heart rate sensor	FM Radio

MP3 Player	RFID Reader
QWERTY keyboard	Thermometer
Biological Agent Detector	802.11a, b, g (WiFi)
Printer	IR transceiver
Alarm clock	Flashlight
Compact Mirror	Calculator
Weight scale	Short range radio (Bluetooth, 802.15.4/ZigBee)

There is even evidence that cameras on phones are not used often and do not provide the compelling user experience the carriers had hoped for [14].

- In the past, there was no way to easily get these devices to collaborate except through physical integration. However, with the deployment of small, power efficient Bluetooth transceivers, this is changing. As Bluetooth chips approach $5.00, it is probable that most devices will incorporate short range, dynamic wireless communication capabilities (most cell phones already include Bluetooth). In this case, collaboration among devices would not require physical integration.

- Almost without exception throughout the history of technology, increased functionality has been accompanied with increased visible complexity. According to Donald Norman, it is this visible complexity that makes devices difficult to use [15]. The most glaring example of this is the personal computer [16]. It is the high level of visible complexity of PCs that causes people so much difficulty and keeps many from using one altogether. The inability to create inherently complex devices with a large number of functions and little visible complexity does not bode well for densely integrated wireless devices.

- Finally, the technology underlying many of these devices (digital cameras, mp3 players, displays) is currently evolving rapidly. Indeed, it is not uncommon for new models of a cell phone to be introduced every six months. This, coupled with the predefined integrated feature set that may not meet a person's needs exactly, creates reluctance on the part of the consumer to purchase the product since a new model with an improved component is coming out in six months, and she can't get the improved component except by purchasing a new model of the integrated device.

On the other hand, a distributed system composed of many separate, collaborating parts has issues of its own. The more separate elements a system has, the more devices to put on, configure, and maintain. Our design principles for a mainstream wearable system must address these issues if the system is to provide the flexibility and incremental growth of a distributed system. These issues are the subject of the next chapter.

3.4.2 A Halfway Point: Modular Systems

A modular system consists of a core device and peripherals. The peripherals physically attach to the core device at specific coupling points. The device accepts different pieces that attach to it that give it new capabilities, form factor, and interfaces.

One example is the Maestro in Figure 3-11 [17]. Maestro represents how short range wireless communication can impact the design of personal communications/computing devices. This modular, adaptable device consists of a base unit that contains most of the intelligence and local applications and acts as the core of the wearable system.

On each side of the base is a removable earpiece that contains a speaker and microphone. Like the display it communicates with the base via

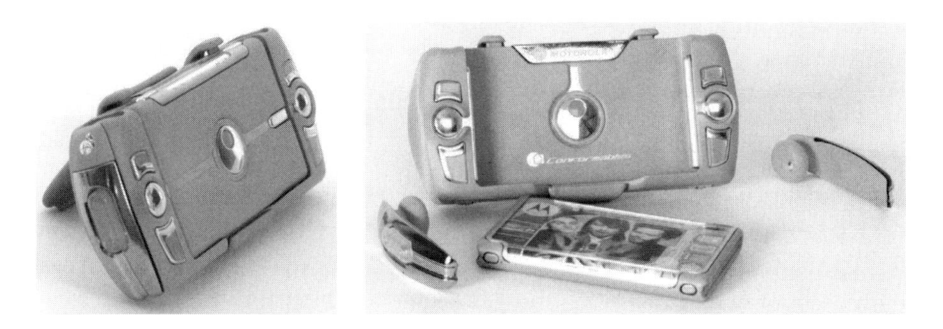

Fig. 3-11. The Maestro Modular Communicator Concept *(Motorola Inc.)*

Bluetooth. When receiving a call, removing either earpiece will answer the call and replacing it back into the base will terminate the call.

The display can be separated and used away from the base. It also communicates with the base via Bluetooth[13]. It contains a touch screen and a high-resolution bitmap. On thc back it contains a digital camera and uses the display as a viewfinder and to preview the pictures.

If modular systems are to be a viable approach to mainstream wearable systems, they must reflect some unique design principles:

- Applications should be aware of the modularity mechanisms as little as possible and not depend on any specific mechanism.
- The modularity options, singularly or in combination, should not impair any feature of the device

[13] Bluetooth EDR (Enhanced Data Rate) provides a 3 Mbps raw channel rate compared with Bluetooth 1.1 raw channel rate of 1 Mbps [18]. The rate, coupled with good compression, could support most video transfer needs between the undocked display and the base unit.

- Peripherals should be easy to dock and undock, preferably without having to look at them
- The modularity options, singularly or in combination, in and of themselves, should not increase the visible complexity of the device

Modularity is an improvement in that it allows the user to attach those devices that are best suited for the task at hand. However, since the extra devices are attached, the bulk of the combined device can grow more rapidly. In addition, the mechanical modularity connectors are often a weak point in the combined device's structure. Finally, there will be limit to the number of devices that can be simultaneously attached to the core device, giving us the same problem of the integrated device. Nevertheless, well designed modular devices can offer a viable alternative to integrated devices.

3.4.3 Distributed Systems

The ideal device is not a device. It is a distributed system composed of devices that are loosely coupled and tightly integrated (from a user's experience sense) with each other. Distributed systems have the following characteristics:

- The system is composed of a central base unit and an arbitrary number of collaborating components. The type and number of components is not fixed and the user can change it to fit the needs of the moment. The base unit would contain much of the intelligence and frequently used applications. It would also have dynamic software adaptability, acquiring new functions and capabilities in response to the changing environment of the user. This is more dynamic than simply downloading applications upon demand. The device itself would sense changes in the user's context and automatically download and install new capabilities. When the functionality was no longer needed, the device could remove the software automatically. This capability depends on a high speed wide area communications infrastructure. Such infrastructures like WiMax [19] are beginning to emerge.
- Each component communicates with the central base unit and can collaborate with other components. The base unit coordinates the collaboration to leverage off of the capabilities of the components and

provide new services. The base unit collaborates with the devices through a short range wireless network.

- The component devices are loosely coupled with the base unit. Each device provides its services with minimal outside visibility and collaborates with the base unit through stable, well defined interfaces.
- New devices and improved versions of current devices can be added to the system and devices removed from it without affecting the operation of the base unit or other collaborating devices.

However, to make the system as a whole easy to use, additional characteristics are required:

- Each collaborating device must itself be very easy to set up for use, be trivial to use, and have a form factor that is unobtrusive when worn or carried.
- There must be a standard framework for designing dialogs between the user and the devices. This is crucial since we want to maximize the flexibility of the system and the ability of the user to rapidly change the system's configuration by adding or removing devices from the system. At the same time, we want to minimize the time the user must spend to regain familiarity with the system's new configuration. This reduces setup effort at the system level and learning effort at the device level.

3.4.4 eClothing: Embedding Wearables into Garments

One of the emerging form factors for wearable systems is clothing. Several examples exist of clothing augmented with communication, computing, and/or sensing devices. Many people have asserted that augmented clothing, that is, clothing that holds or embeds electronics, is inevitable and at some time in the near future most everyone will be wearing it. However, to date these efforts have not caught on in a major way. Let's look at why.

There is a spectrum of integration of wearables with clothing (see Figure 3-12). At the lowest level of embeddedness is a Packed configuration in which discrete devices are hung on the outside of the garment or inserted into outside pockets. The devices are separate and self contained. The garment contains no power or data infrastructure to enable the devices to collaborate with each other or with the user although it may provide guides

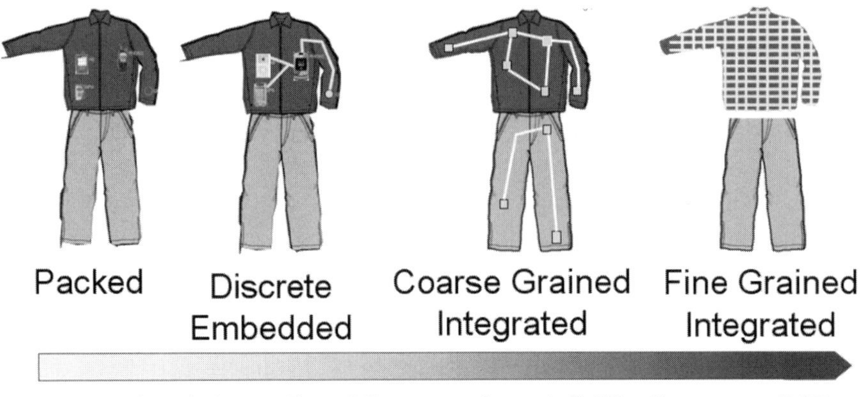

Fig. 3-12. Spectrum of Embeddedness

for routing earpiece and microphone wires from a device to another point on the garment such as the collar. Beyond increasing the size of the pockets to accommodate electronic devices, the garment provides no support for the installation of devices. An example of this level of embeddedness is the Scott eVest we saw earlier (Figure 3-2 left).

Beyond that is a level we term Discrete Embedded. In a Discrete Embedded configuration, the garment provides internal pockets or securing mechanisms specifically designed for incorporating electronic devices in the garment. The garment may also provide support for powering and/or controlling the devices. This is the state of most garments today and an example is the ICD+ jacket (Figure 3-2 second from the left).

The next step is Coarse Grained Integrated. Here the electronic devices are split apart into their major components. These components are distributed within the garment. For example, a cell phone would have its front and back covers removed. Its RF board and other circuit boards would be integrated into the inside liner of the jacket. Its user interface would be placed on the jacket's sleeve. The devices are disaggregated in order to achieve a less obtrusive fit within the garment. The Media Lab Mithril vest (the 2 rightmost images in Figure 3-2) is an example of Course Grain Integrated.

The Coarse Grained Integrated configuration requires the garment to provide an infrastructure to bring power to the components. A data bus is also required to allow the different devices and components of a device to communicate. In addition, a wearable system controller is required to manage and coordinate the collaboration among the devices. Each device would be responsible for managing communication among its own components.

At the high end of the wearability embeddedness spectrum is Fine Grained Integrated. Here devices and device components no longer have their own identity. Instead, the garment contains its own communication and computing network, complete with power and data distribution infrastructure. The electronics are totally integrated into the garment. The user interface consists of speech with a display and possibly a keypad embedded in the garment's surface. There is currently no example of this level of electronic embeddeness in a garment.

As we move up the levels of embeddedness the permanence of the device embedding increases. For example, devices are easily inserted and removed in a packed level garment such as the Scott eVest. However, by the time we reach a garment at the Course Grained Integrated level, such as the MIThril vest, the electronic devices must be considered permanently embedded in the garment.

While most of the examples of augmented clothing have come from research labs, there have been some commercial products. Almost all have failed commercially. One of the earliest products was the ICD+ jacket [2] shown in Figure 3-2[14]. Starting in 1997 a multi-disciplinary team at Philips, comprised of experts in electronics, consumer design and fashion,

[14] ICD stands for Industrial Clothing Division.

researched electronic fabrics and clothing. In late 2001Philips announced the ICD+ jacket based on this research. It was a joint project between Philips and Levi-Strauss & Co researchers in San Francisco.

Four different jackets styles were designed. Each contained an electronic network of almost four feet of wires woven into the jacket that connected the devices. This network allowed the control of a Philips GSM mobile phone and an MP3 player (both included with the jacket) via a common remote control in the jacket. The control device contained a small display that indicated an incoming phone call, e-mail, or the title of a song playing on the MP3 player. Headphones and a microphone were incorporated into the collar. The earphones were stored in rubber housings below the collar on the front of the jacket.

The jacket was sold only in Europe and cost US$600 to US$900 depending on the style. Only a couple thousand were made and they are no longer available.

Other jackets with embedded electronics have appeared, for example the recently released Burton – Motorola Audex jacket [21]. Announced at the 2005 Consumer Electronics Show, the jacket went on sale in late 2006. The jacket uses Bluetooth to link the user's cell phone and iPod with a removable control module on the jacket sleeve. The jacket's hood contains integrated stereo speakers and a microphone near the collar. This is another example of Discrete Embedded electronics. In 2007 the jackets sold for up to US$649.95 (including all internal electronic components but not including the iPod or cell phone) and come with a long user manual.

There have been other form factors involving clothing, mostly in the research labs and universities. The MIThril Vest [3] mentioned above consisted of a vest which held the devices, wiring, battery, and supporting electronics (Figure 3-13, left). Specific pieces of hardware were secured to the vest via Velcro attachment points. The vest itself zipped into an outer covering, much like a liner in a jacket (Figure 3-13, right).

The vest contained a network of data and power lines which linked of both repackaged, off-the-shelf hardware and custom components. The hardware, together with the network, formed a distributed wearable system

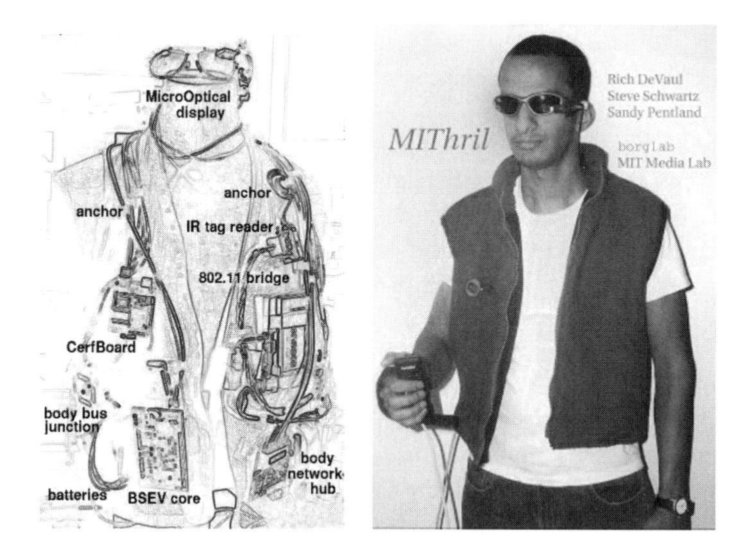

Fig. 3-13. MIThril Schematic Showing Major Elements and Inserted in the Outergarment *(MIT Media Lab / Alex Pentland)*

consisting of processors, storage, wireless transceivers, and a range of sensors.

A single set of batteries powers the devices, using the vest's power line network. This significantly cuts down on the maintenance of device power sources and results in lower total battery weight.

Another interesting example of electronics embedded in a garment is the SmartShirt from Georgia Tech [22]. The SmartShirt, also called the Wearable Motherboard™, provides a platform on which sensors and computing elements are attached (see Figure 3-14). Originally developed under contract with the US Department of the Navy, the shirt had an uninterrupted length of fiber optic tubing going around the shirt from top to bottom.

Light travels from one end of the plastic optical fiber to a receiver at the other end. If something (such as a bullet) penetrates the shirt, it will likely break the fiber and interrupt the beam. The beam reflects back to a receiver at the beginning of the fiber and helps to determine where the shirt was penetrated. This can be radioed back to medical personnel who can determine the location of the soldier's wound and assess its severity.

The shirt also has an embedded data grid that transmits information between sensors and a controller, also connected to the shirt. The sensors plug into connectors similar to “button snaps” used in clothes. The user can attach sensors at almost any location on the shirt using these snaps. This creates a flexible “bus” structure that can accommodate people of various sizes.

One of the most significant aspects of the SmartShirt is that the optical fiber and data grid are integrated into the structure during the fabric production process. There are no discontinuities at the armhole or the seams, due to a novel modification in the weaving process [20].

In 2000 Sensatex was founded to commercialize the SmartShirt. On May 1, 2007 the company announced that they would begin field trials.

With very few exceptions, none of the ‘eGarments’, including those discussed above, have been great commercial successes[15]. Why is this? One obvious reason may be price. Most of these garments or jackets have very high prices. Recall that the ICD+ started at US$600 and the Audex Cargo jackets current sell for US$649.95 [23]. That is a lot to pay for hands free listening of music and hands free conversations on your cell phone.

However, there are other reasons, some of them having to do with the nature of clothing and electronics. Clothing is all about image, feelings, comfort. Electronics is cold, logical, and stiff. The merger of the two seems at first incongruous.

[15] The Audex jackets from Motorola have not been on the market long enough to determine whether they will be a success or not.

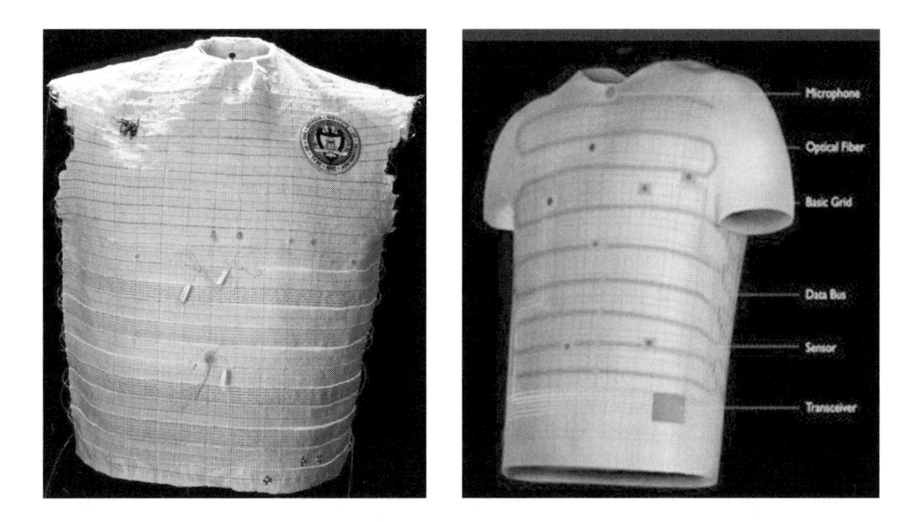

Fig. 3-14. The SmartShirt Prototype (left) and final concept (right) *(Courtesy: Textile Information Systems Research Laboratory, Georgia Institute of Technology, Atlanta, Georgia, USA)*

If a wearable system is to be embedded in a garment and that garment is to be widely accepted by the mainstream population, the focus must be on the garment, not the wearable system. For example, if a wearable system is embedded into a jacket, the user must be able to think of and use the enhanced jacket as a regular jacket; that is, as if the jacket did not have the embedded electronics. Using the enhanced capabilities should not require actions or mindsets that are incompatible with the accepted concept of the unenhanced jacket. Requiring a long, detailed process to remove the electronics before washing the jacket and replacing them before using it afterward is not treating the enhanced jacket as a typical jacket. Neither is buying a jacket and having to read a long user's manual.

This has significant implications for designing garment embedded electronics. One of them is that the electronics must be protected in such a way as to make it unnecessary to remove them before washing, dry cleaning, and ironing. Another is that the embedded network must be fault tolerant. That is, should one path between two embedded devices break due to handling or cleaning the garment, there must be a way to route around the broken network link.

Another possible reason why eGarments have not penetrated the mainstream population is that we change clothes - a lot. We change them with the season, we change them with the activity - we even change them with the time of day. For example, many people change from formal working clothes to more relaxing clothes when they get home.

Unless we are ready to buy several instances of each of the embedded devices so they can remain in the clothing, we will have to remove the devices from one set of clothes and place them into the clothes we are changing into. This requires the devices be easily accessible in the clothing and easy to remove and insert. Even then, this can be very laborious and time consuming, a high price to pay for the benefit.

That is why it is likely that the type of devices that will be successfully (from a user acceptance point of view) embedded into a garment will be those devices and services that enhance or augment the primary function of the host garment or one of the primary activities engaged in while wearing the garment. For example, the devices and services embedded into a ski jacket could include an outside air thermometer, moisture sensor, safety beacon, and sensors to detect if the user is standing or lying down. All of these sensors support one of the primary activities being performed while wearing the jacket: skiing.

This means there is less reason to remove the sensors in the jacket when moving on to another activity since a different set of sensors is likely employed for that other activity. In any case, the general purpose functions such as communications, context awareness data fusion, etc, are kept in the Wearable System Controller (WSC), a device that is small, easily transported from one garment, belt, or purse to another, and transparently interfaces with the specialized devices and applications in the garments being worn by the user.

3.5 TIME OUT: CONVERSATION WITH A SKEPTIC

Thomas is a skeptic, and as such, asks tough, pointed questions about wearables. Thomas is skeptical about the whole idea of wearables and whether people will ever accept this technology. This conversation previews many of the ideas and issues addressed in the rest of the book.

Thomas: I have a hard time buying into this whole wearables thing. Why would people ever trade in their cell phones, PDAs, and PCs for a wearable?

JD: PCs, PDAs, and cell phones are just tools. It is very rare that you want to use a phone or PC for its own sake. Rather, you use it because it is currently the best way to perform the task that is important to you. In the case of the cell phone, it may be to talk to someone who is too far away to do so in person. In the case of the PC, it may be to write a letter or record business expenses. The point is, you are using the phone or PC only because they are currently the easiest or only way to perform your true task.

However, too often, our tools get in the way of what we are trying to do. How many times have you had to wrestle with a word processor to get it to format something the way you wanted it? How often have you juggled what you were carrying when your cell phone rang so you could retrieve it and answer the call? So the real issue is: can wearable technology offer tools that interpose themselves less between the user and their real task than current tools such as the PC and cell phone?

Thomas: Are you trying to tell me that the phone will eventually disappear? I just don't buy that. I think the phone will always be around.

JD: The phone as a standalone device will not disappear – at least not any time soon. However, it is likely that, in its current form, it will not be the most common tool for mobile communication in 5 – 10 years.

As computing and communication become embedded into most objects around us – a trend that has already started – there will simply be so many different devices with which we can interact. Many of these devices will have form factors that allow them to be used in specific situations where they will be much easier to use than the current cell phone.

Thomas: Look, they are building very cool phones with cameras, MP3 players, large screens, and FM radios in them. Why can't they just put everything you need into the phone? After all isn't technology always getting more powerful and electronics getting smaller?

JD: There is no denying the attractiveness of a single device. This would eliminate putting on multiple devices and making sure the devices were properly charged, and maintained. However, it is unlikely that a single device, however adaptable, will ever be suitable for all situations.

The ideal device is not a device. It is a system composed of devices that are loosely coupled and tightly integrated (from a user's experience sense) with each other. The central unit would have dynamic software adaptability. It would also have the hardware adaptability, but would achieve it logically with a wireless PAN, rather than through physical integration.

Thomas: But now I have to buy multiple devices. Won't that in the long run cost me more money?

JD: Perhaps, perhaps not. Each of these devices will be much simpler since it can leverage off of the processing, storage, and applications of the central unit. So they may be less expensive. Also, you will have a much wider variety of devices from which to choose so you can buy devices with the functionality that most effectively meet your needs rather than settling for the inherent compromises that accompany each element of an integrated device. And you will be able to buy only those devices that you want instead of paying for all of the functionality in an integrated device, most of which you may not want or use. After all, most people use only a small percentage of the functions in their cell phones.

Thomas: I suppose you believe these systems will eventually replace the PC too.

JD: Not at all. A wearable system is not a replacement for the desktop or laptop PC. It has a different focus. The PC is mainly focused on tasks involving elements outside of us and of a somewhat impersonal nature (media files, spreadsheets, presentations, etc). The wearable system, on the other hand, is more focused on the person: on us, on understanding where we are, what we are doing, who we are with and so on. They will be smart about us.

Thomas: OK, what kind of killer applications will we be using?

JD: Whole new classes of compelling applications become possible. For example, reminders that are conditioned on your situation, not just time and date. So you could say "Remind me at 10 a.m. tomorrow if I am at work and need a ride home to call Tom". Imagine being able to control almost any device by speech – even those without speech recognition capabilities. Or imagine that your system always knows where you are and can determine how to best notify you of incoming information based upon your current location.

Thomas: Whoa, wait a minute. I'm not sure I want everyone to know where I have been.

JD: In many cases you may choose to have your location information kept on your wearable system and not sent to the location service in the infrastructure such as a web server. The point is there will be choices you can make regarding how you share your location and other private information.

Thomas: Ok. So say I buy into this idea of a wearable system instead of an integrated device. I still don't want to have all of these devices hanging off my belt or my clothes. It would ruin the look. After all, a person must maintain a proper image. Heck, I'd bet even geeks care about how they look at some level.

JD: No comment on that last point. But you are right about the belt thing. We do not want people wearing all kinds of devices on their belt like Batman.

These devices don't just hang on you. Rather, they conform to you – to the shape of your body, the movement of your body in all planes, and orientation of your body in all postures. And some of these devices could be embedded within your clothing, further reducing their obtrusiveness.

Thomas: Yeah, I read about these so-called "smart" jackets and shirts. Somehow, the thought of clothes having all these circuits and electronics in them just leaves me cold.

JD: Many of these smart garments now being prototyped do not reflect a lot of effort to make them fashionable. Many are simply engineering prototypes that the news media has gotten a hold of. In the smart garments that have been released as products, for example the ICD+ jacket from Levis and the Audex jacket from Burton, devices are simply attached to special connectors in pockets. There is not a great deal of what we call fine-grained embeddedness in which the devices are disaggregated and distributed throughout the garment. That would make the embedded devices much less noticeable. However, it makes the design of the garment much more complex.

Thomas: Ok, but I have enough stuff now for which I have to charge batteries. I just don't see how people will accept having to take off all these gizmos and charge them every day and then put them back on. And what happens when I want to change clothes? It's just not worth it.

JD: You're right. Minimizing the maintenance of these embedded electronics within the garment is one of the biggest challenges. That is why devices that are embedded in garments will most likely be specialized devices that reflect and enhance the specific function of the garment. For example, a ski jacket may have embedded sensors that detect and report the air temperature, wind speed and figure out the wind chill factor. If it gets too cold, it may remind you of this. The more general purpose functions and applications, including the speech user interface and those applications you want with you all the time will be implemented in the system's central unit, a small unobtrusive device that you transfer from outfit to outfit.

As for those devices actually embedded in clothing, what you do not want to do is completely change how you deal with the garment. If it is a shirt for example, you will want to deal with it as a shirt. So we must find some way of maintaining the devices within the normal usage patterns of the garment.

One example is to charge the electronics in the shirt while it is on the hanger. And do this just by putting the shirt on the hanger, nothing more. So you charge the electronics just by following the care patterns of the shirt itself. We must find ways to do this for all aspects of embedded electronics.

Thomas: But suppose I still want to read email on a wearable system.

JD: You will actually have many choices about how to do that and can select the one that best fits the environment you are currently in. For example, you could walk up to any monitor that supports connectivity with your wearable system and use it as a display. You could pick up any kind and size of keyboard and use it.

Alternatively, if you are out and about and not near these devices, you could put on your display glasses and see a virtual screen suspended about 18 inches in front of you. You could use your speech interface for input.

Thomas: I can't imagine walking down the street while I am looking at a screen suspended in midair in front of my face. That is a great way to walk into a pole!

JD: I agree, this is an issue. There is obviously a responsibility to use these devices in the appropriate situation. So if you do use your display glasses, you may want to sit down so that you do not walk into someone or something. If you cannot stop and sit down, you use only the speech interface that provides automatically summarized versions of the email. The wearable system adapts to your situation.

Thomas: So how important will speech be in these systems? I've tried using those speech dictation systems on the market. Sure, most of the words you say its gets right. But about every 10th one you have to correct. What a pain!

JD: The systems you speak of are dictation systems. They allow almost unconstrained vocabulary and natural speech. This is currently extremely difficult and requires significant computing resources. Most wearable systems will use some variation of command and control that use grammars to specify what exactly can be said. The user must speak the utterances specified in the grammar or the system will have little chance of recognition.

Thomas: Come on, you can't expect people to learn to speak a certain way when using these systems!

JD: There are precedents for this. For example, people have readily adopted Graffiti, the stylized printing used by the Palm character recognizer. People also adapted to the need of the first speech recognizers that required people to briefly pause between words.

However, the objective of wearable systems is to make the devices as transparent as possible. Therefore, the structure of the grammar phrases must reflect as closely as possible the common speaking patterns and conventions of those who will use it.

Thomas: Ok, Perhaps speech will work all right. But what happens when I am in a meeting or in the library? I can just see it now: Librarians rushing around frantically, shushing everyone to be quiet and stop talking to their wearables.

JD: Speech is never the only interface. Other interfaces such as text, GUI, haptic, and gesture must also be available. The goal is to allow the user to employ the interface best suited for the current context. In addition, the user must be able to switch from one interface to another seamlessly. So while you might be using the speech interface while driving to the library, you would easily switch to a text or GUI interface when entering the library.

Thomas: Yeah, that does sound neat. But if you think I'm going to walk around, talking into thin air, forget it. I have enough people already who wonder if I'm wacko.

JD: There are two answers to this. The first is that as wearable systems become more common, people will become more comfortable speaking into the air when they use the speech interface or are talking to someone else using a hands-free and eyes-free communications device. People are already using earbud speakers and string microphones and wireless Bluetooth headsets with today's cellular phones, enabling hands and eyes free communication. While people were at first startled at this, it has become common enough that people quickly figure out what is going on and don't give it another thought.

The other answer is that we nevertheless must take this and other social conventions and taboos into account when designing wearable systems. Creating designs that provide subtle but effective visual or audio cues about what the user is really doing will go a long way to removing some of the concerns and resistance people will initially feel toward these systems.

Thomas: Alright, I'll think about it. (Long pause). So, um when can I get one?

REFERENCES

[1] SCOTTEVEST [SeV] - Technology Enabled Clothing, http://www.scottevest.com/index.shtml, accessed March 20, 2006.

[2] Peterson, K. F., Now You Can Have Your Phone ...and Wear It, Too, 10 meters.com Industry, http://www.10meters.com/levi_jackets.html, accessed March 20, 2006.

[3] MIThril, http://www.media.mit.edu/wearables/mithril/index.html, (October, 2003), accessed 2/14/06.

[4] Stanford V., 2002, "Wearable Computing Goes Live in Industry," IEEE Pervasive Computing, vol. 01 p. 17

[5] Lein, A. Z., 2004, Nokia Medallion I, PocketNow.com, (March 1, 2004), http://www.pocketnow.com/index.php?a=portal_detail&t=reviews&id=336, accessed March 20, 2006.

[6] Maxim Integrated Products, 2006, What is an iButton?, http://www.maxim-ic.com/products/ibutton/ibuttons/, accessed March 20, 2006.

[7] Krantz M., The Ubiquitous Chip, Time.com, December 29, 1997 / January 5, 1998 Vol. 150 No. 28, http://www.time.com/time/moy/daily3.html

[8] Mann, S, 1998, Wearable computing as means for personal empowerment, Keynote Address for The First International Conference on Wearable Computing, ICWC-98.

[9] Twiddler2, 2006, Handykey Corp., http://www.handykey.com/site/twiddler2.html, accessed March 20, 2006.

[10] Mann S., (1998), Humanistic Computing: "WearComp" as a New Framework and Application for Intelligent Signal Processing PROCEEDINGS OF THE IEEE, VOL. 86, NO. 11, NOVEMBER 1998 pp 2123 - 2153

[11] Farmer, Eric, Chapman, V., Brownson, A., et al., 2003, Report on Work Package 3: Prevention of Performance Impairment, CARE Innovative Action Cognitive Streaming Project. 13.

[12] Mann S., Intelligent Image Processing, John Wiley and Sons, 384pp, November 2, 2001, ISBN: 0-471-40637-6.

[13] Proceedings of the IEEE, Vol. 86, No. 11, November, 1998, Pages 2123-2151

[14] In-Stat, 2006, Disappointment with quality and cost limits usage of camera phones, Scottsdale, Ariz., (February 6, 2006), http://www.instat.com/press.asp?ID=1574&sku=IN0502053MCD, accessed March 21, 2006.

[15] Norman, D., The Design of Everyday Things, Currency Doubleday, New York, 1988, pp. 99-102.

[16] Norman, D., The Invisible Computer, The MIT Press, Cambridge, MA, 1998, pp 77-78.

[17] Ferenczi, P. M.,2006, Super Prototypes, Laptop Magazine. 24(): 98-99.

[18] McCall D., 2006, Taking a Walk Inside Bluetooth EDR, http://www.commsdesign.com/design_corner/showArticle.jhtml?articleID=55800768

[19] WiMax Forum, http://www.wimaxforum.org/home/

[20] Cascio J., 2004, Weaving the future, (November 9, 2004), http://www.worldchanging.com/archives/001555.html.

[21] Motorola, Audex Home, http://direct.motorola.com/hellomoto/audex/default.asp, accessed February 14, 2006.

[22] Georgia Tech Wearable Motherboard™, 2002, http://www.smartshirt.gatech.edu/, accessed February 14, 2006.

[23] Burton online store, accessed 11/9/06 http://store.burton.com/Collections/Audex.jsp?bmUID=1163075277585

PART 2: MAINSTREAM WEARABLE DESIGN

Chapter 4

OVERVIEW OF MAINSTREAM WEARABLE DESIGN

Operational: adj. Of or relating to an operation or a series of operations.
Inertia n. Resistance or disinclination to motion, action, or change

4.1 WHY THE BAD EXPERIENCE WITH ELECTRONIC DEVICES?

When was the last time you curled up with a PC and really experienced its start up process; feeling a sense of excitement as the startup progress bar moves toward the other end; relishing the opportunity to have some down time while the machine loads and configures your setup options. And, as the startup process nears its conclusion, basking in the richness of its startup screen with its solid blue (or green) background or soothing image. You reflect that the experience was well worth the two to three minutes delay before it became ready to do what you had in mind in the first place.

Or remember when you appreciated the opportunity to test your eyesight by trying to read a web page with blue text on a black page background (see Figure 4-1). The chance to exercise your eyes was surely worth the effort and delay it imposed on trying to read the article.

And finally, I'm sure you remember when you proudly carried your first cell phone. It was easy to impress your friends since they were sure to see it on you, given its size and bulk. The discomfort and inconvenience getting in and out of the car while wearing the phone on your belt was definitely offset by the status it afforded you. Plus, the fact that it kept constraining your movement constantly reminded you that you had such a marvelous device.

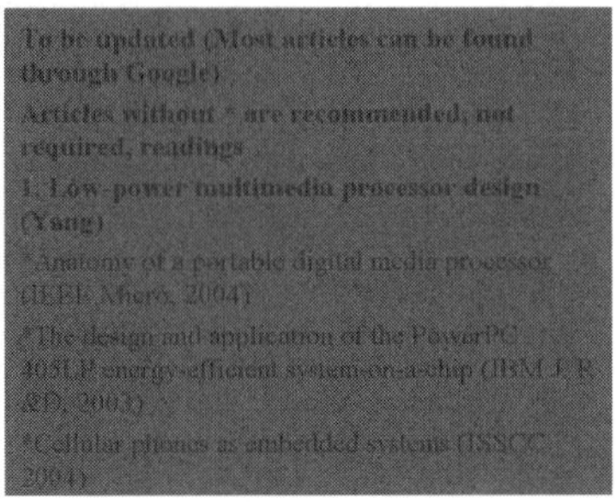

To be updated (Most articles can be found through Google)
Articles without * are recommended, not required, readings
1. Low-power multimedia processor design (Yang)
*Anatomy of a portable digital media processor (IEEE Micro, 2004)
*The design and application of the PowerPC 405LP energy-efficient system-on-a-chip (IBM J. R &D, 2003)
*Cellular phones as embedded systems (ISSCC 2004)

Fig. 4-1. Difficult to read text

You miss all that with these new phones that are so small that you can hardly tell you are wearing them since they cause you no discomfort.

Yeah, right.

The truth is, you never appreciated any of these experiences. With the possible exception of the very first time they occurred, they were to you unnecessary inconveniences, if not sources of outright frustration.

What each of these situations has in common is that in each you had a specific task you wanted to accomplish (your primary task), such as reading an email, or ordering a product, or being able to talk to a person separated from you by a distance that made face to face conversation impossible.

However, in each of these situations, your primary task had nothing to do with these devices. The devices were simply the mechanism you chose or you were required to employ to perform the task [1]. That is, you did not want to use the device for the sake of using and enjoying the device itself.

Yet in each case the device (PC, cell phone) or service (web page) made you take notice of it. It made you switch your attention from your real task to the task of getting the PC ready for use, compensating for the deficiencies in the design of the Web page, and accepting the constraints and discomfort as you went about your daily activities wearing the phone in order to have it there when you needed it.

4.2 A TRANSPARENT USE DESIGN MINDSET

Why were the experiences above so unsatisfactory? Each of those devices or services had too much ***Operational Inertia***. Operational Inertia is a fundamental concept that is crucial to the design of mainstream wearable devices, services, and systems. It is defined as

> ***Operational Inertia is the resistance a device, service, or system imposes against its use due to the way it is designed.***

Operational Inertial is composed of three elements: setup effort, interaction complexity, and non-use obtrusiveness. Each of the above scenarios illustrates one of these elements.

1. **Setup Effort**: Setup Effort is the time and effort required to bring a device, service, or system to the point where it is ready to be used for its intended purpose. In the first experience above, the time and effort required for the PC to boot up and configure itself so it is ready for use is its setup effort. It also includes how often the same operation must be performed, for example giving your telephone number every time you are transferred to someone else on a customer support help line.

   In the case of a phone, it includes all of the operations we must do with it before we are actually talking to the other party. This includes retrieving the phone from its holster, pocket, or purse, opening the flip and/or extending the antenna, looking up a number, dialing it, and waiting to be connected.

   Setup effort rarely, if ever, has anything to do with the real task we are trying to accomplish. It is simply the effort we must go through to ready the mechanism we are employing to get our real task done. While the setup time and effort may be critical to the device, from the standpoint of our original task, it is irrelevant and a waste of our time. Clearly a device that had little or no setup effort would be a more attractive choice to get our task completed.

2. **Interaction Complexity**: Interaction Complexity is the measure of how difficult a device, service, or system is to use once it has been set up for its intended use. This includes how difficult it is to remember or give commands and how complex an application's mental model is. It also includes how difficult a device is to manipulate and control.

Examples with using a PC are deciphering confusing messages when trying to recover from errors, follow confusing installation instructions, and figuring out why an application took unexpected action. Interaction complexity for a web site includes trying to read text with low contrast from its background (ex., dark blue on black), dealing with popup ads that obscure the page text, and encountering and recovering from broken links. For cell phones as they get smaller, their buttons shrink and screen text becomes harder to see. This results in increased interaction complexity with the phone.

Interaction complexity is one of the most difficult elements of Operational Inertia to reduce since it is so closely associated with the user's actions in the application. The more closely the application's actions and responses reflect the user's mental model of the task they are trying to perform, the lower the Interaction Complexity. All actions required by the application that do not follow this mapping of the user's expectations to the actions of the application are, from the user's task point of view, irrelevant, a waste of time, and a prime source of frustration[16].

3. **Non-use Obtrusiveness**: Non-use Obtrusiveness is the measure of how often the device, service, or system makes us uncomfortable or aware of its presence when we are not using it for our primary task. (The obtrusiveness of a device while it is being used is part of Interaction Complexity).

   For example, how often does the device constrain my motion or cause me discomfort as I move or assume different postures? In other words, how often does it remind me of its presence when I am not using it or do not feel it is necessary to be used? As an example of the latter, consider a service that keeps interrupting me with information I cannot use or in

[16] Norman [2] calls this mental model to action and response mapping 'the Gulf of Execution'. The width of the gulf reflects the distance between the user's mental model of the object's operation and the object's actual actions and responses.

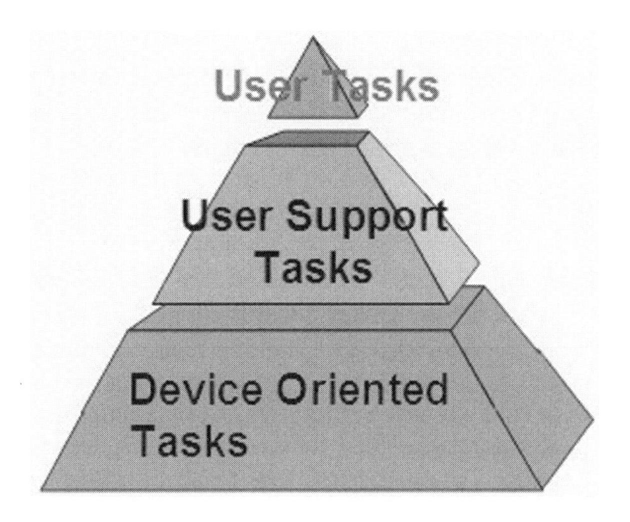

Fig. 4-2. User Task Hierarchy

situations where the method of notification is inappropriate (for example public audio in a library). High levels of non-use obtrusiveness significantly reduce the user's desire to wear and/or use the system, greatly reducing its effectiveness.

Non-use obtrusiveness can be experienced in surprising ways. For a PC, it includes the desk space it takes up. To the extent that this reduces the desk space available to do other work not requiring the computer, it is a sign of the PC's non-use obtrusiveness. Non-use obtrusiveness for a web page includes hearing sounds made by ads on a page when the page is not in the foreground and you are not using it and popups activated by the page when the page window is in the background and you are not using it. For a cell phone, non-use obtrusiveness includes it getting snared by the seat belt when you are getting out of the car while wearing it on your belt and it getting in your way when you twist or bend.

4.2.1 The Primary Task

The term 'primary task' has come up several times and we need to define it. We categorize a task into one of three classes based on its relevance to user's goals. Figure 4-2 shows this task hierarchy.

1. **Primary Task** This is a task of direct relevance to the user's goals, the "real" task. Experience with these tasks forms bulk of user opinion about the performance of a system. Examples are communicating with someone, purchasing a product, or taking medicine. Since the task is directly relevant to the user's goal, the user is willing to expend some effort and attention on it.

2. **Support Task** This is a task of limited user interest but with visible user benefit and primary task relevance. Because the user can see the benefit to the primary task, they may not be too upset if they experience brief side affects of the task, such as decreased system performance while the task is running.

   An example is an email system checking a message for viruses. While the system is performing this task the user may experience a delay in seeing the message. However, since they understand that if this is not done, they could eventually suffer serious damage to their computer and files from a virus attack, they will accept the reduced system performance without too much annoyance. Another example is when you have to reenter your credit card information when it changes. Without doing so, you will not be able to purchase the item you are there for, nor any items in the future from that site. So you take the time to do this without too much grumbling.

3. **Device Task** These are tasks relevant only to the operation of the device or service. The user sees little or no relevance to their primary task. Have you ever been working on your PC and all of the sudden there is hard disk activity and the system begins to run slowly for a period of time? The PC is performing low level system oriented tasks such as paging idle programs out of RAM and into virtual memory, increasing the size of the virtual memory partition, reorganizing the hard drive or any number of low level housekeeping chores. However, from your perspective, these tasks have little or nothing directly to do with your

primary tasks, which may be to write a document or read email. Therefore you regard them as a pure annoyance. They are being done solely because of the way the PC hardware and operating system are designed, not because of any task you are doing. So you are much less tolerant of their effect on your primary task.

If wearable systems are to go mainstream and approach transparent use, system designers must not allow device tasks to become visible to the user and must minimize the intrusiveness of support tasks on the user's performance of their primary task.

Another important question is what we mean by 'transparent use'. For the purpose of this book, transparent use implies the user's primary task is accomplished without

- focusing on required devices and/or services;
- extensively manipulating devices and/or services; and
- dealing with the constraints imposed by required devices and/or services.

Note that this does not require 'simple' devices, just that devices appear to user to be simple. Similarly, when we talk about zero or near zero Operational Inertia, we are not talking about devices, services, or systems with zero or even near zero complexity. Wearable systems are complex. More than that, they can be *inherently* complex. That is, there is currently no design or architecture that can make them simple.

However, the element that causes devices, services, and systems to have a lot of Operational Inertia is not complexity. Rather it is *visible* complexity [3]. This is the complexity that is visible to, and experienced by, the user. In all likelihood low Operational Inertia systems will be remarkably complex. However, all that complexity would be hidden from and not experienced by the user if the system is to be transparent to use.

4.2.2 A Closer Look at Operational Inertia

The definition of Operational Inertia refers to devices, services, and systems and differs in its scope and causes in each.

Device Operational Inertia

Operation Inertia is most visible in a device since it involves the physical properties of the device and the user.

Setup Effort: The setup effort for a device involves all those actions the user must perform to get the device ready for its intended use. The actual activities and their sequence depend on where the user is wearing the device and how the device is configured.

As an example, consider the setup effort for a typical cell phone. Depending on how the user has configured the phone, most or all of the following activities must be performed before the phone can be used for its intended purpose:

- Retrieve the phone from where it is being worn or carried. This can be a holster on a belt, in a purse, or in a pocket.
- Orient the phone in space. This typically involves rotating it about one or more axes so that the front of the phone is facing you.
- Open the flip if the phone has one. If the phone has no flip, the user may have to unlock the keypad
- Extend the antenna if the phone has one and the signal strength display indicates poor reception with the antenna retracted
- Look up the number in the phone book if you don't remember it. This can also involve pressing buttons to scroll to the number. If you remember the number, you must enter it manually. As an alternative, if the number has a voice tag and the ambient noise is not too high, you can speak the voice tag after pressing the speech recognition activation button. Once the number is entered or selected, you must press the send button
- Waiting until the call is connected. This can involve a significant amount of time if the called party must be first located within the cellular system. You must wait while the phone rings and the person answers the call.

Only once the last of these activities is completed can you use the phone for its real intended purpose: to carry on a conversation with a person from whom you are separated by a distance that makes face to face communication impossible. All of the activities above, from the standpoint of this purpose, are irrelevant and a waste of time. They exist only because of the way the cell phone and network are designed.[17]

The presence of this setup effort influences how and when you use the cell phone. For example, suppose you and your friend are at the supermarket and you are separated by one aisle. You want to check with your friend on which brand of a product to buy. You have three choices:

1. You can yell loud enough so that your friend can hear you and your friend will yell back the answer so you can hear it.
2. You take out your cell phone and go through the above setup process to call you friend and ask which brand of the product you should buy.
3. You walk the short distance to the other aisle and discuss with your friend face to face which brand of product to buy and then walk back.

In all likelihood you will follow the third course of action and walk the short distance to discuss the issue with your friend face to face and then walk back and select the item. Why? Well, it is considered impolite and can be embarrassing to yell as in the first option. And it is just not worth the effort to follow the second course and make the phone call (unless the act of walking is difficult for you, for example if you are on crutches or elderly).

However, if you could eliminate all setup effort of using the phone, that is, if you could simply look at the shelf and speak, without manipulating anything, in a normal tone and volume, "Tom, which do I buy, brand A or Brand B?", and Tom continues what he is doing without any change and simply answers, in a normal tone and volume "Better buy Brand B", you may be more inclined to use the cell phone and stay by the shelf. This is

[17] Users of a Nextel phone's Push To Talk service reduce the setup time by virtually eliminating the time required to set up the connection through the network (the last setup task). This time reduction can be significant which accounts for the loyalty of most Nextel customers. However, the user must still perform all or most of the other setup tasks.

because the effort required to do this is low enough that you consider it acceptable to do in this 'trivial' task.

The above supermarket scenario is an example of *Opportunistic Communication.* Opportunistic communication is defined as 'communication that is performed only because it is nearly effortless to do so'. Throughout our daily activities we encounter many opportunities for communication. These involve communication with people who, although close by typical cell phone calling distances, nevertheless are too far for face to face communication without yelling. Other examples are moving apart in a house while talking (the Speaker Tracking application in Chapter 2), speaking to occupants of the car next to yours, and conversing with a friend sitting four aisles from you on an airplane.

Setup effort also includes maintenance. One of the most serious maintenance issues for wearables is that of recharging the device power sources. Note that switching to new power sources such as fuel cells may reduce, but will not eliminate this problem. As the number of wearable devices in the system increases, so does the seriousness of this issue. When wearable devices are embedded in clothing this can be especially difficult to address since the power sources may not be readily accessible and thus may involve a series of steps to access the batteries and charge them.

Interaction Complexity: The interaction complexity of a device is the difficulty in using the device's physical affordances[18], independent of any difficulties imposed by the specific application, once the device is completely set up for its intended purpose.

Interaction complexity also includes how difficult the device is to handle and manipulate while using it. We call this 'in use obtrusiveness' (to

[18] An affordance is a perceived or actual property of a device that determines how it can be used [4]. For example, some affordances of a PDA are its stylus, touch screen, buttons, and display.

differentiate it from nonuse obtrusiveness discussed next). The more difficult the device is to handle while using it, the more you have to focus on the device and the less you can focus on your primary task,

Interaction complexity involves the difficulty in using the core elements of the device. This can include how difficult it is to use the menus for the core function (functions common to all applications on the device), the difficulty in getting help, and constraints on your behavior imposed by the device during use.

An example of the latter can be found in many car radios. Most car radios today turn off immediately when you turn off the engine. Now, when you want to listen to the car radio without the engine running, you must turn the ignition from 'on' to 'acc' to power the radio from the battery and turn off the engine. During this transition the radio looses power for a moment and turns off. It turns back on when the key reaches 'acc'. Depending on the design of the radio, it can then take a couple of seconds to restore the audio. This interruption can be irritating.

However, many new cars retain power until you open the door with the engine off. The result: the ability to transition to listening to the radio while the engine is off without interruption. A small matter perhaps. However, designs can have unintended consequences. How many times have you not turned off the engine because you did not want to interrupt what you were listening to? The result: increased gas consumption and air pollution.

One of the principal determiners of interaction complexity for a device is the characteristics of the *mental model* of operation applied by the device's designers. A mental model is simply the concept of the device's operation the designers had in mind when developing the device. This model governed the decisions they made about the way various elements of the device interact with each other and with the user. It influenced how deep the menu hierarchy is, how you ask for and receive help, and the structure of the speech commands you use. It also influenced how the controls are laid out, how icons are designed, and whether or not you can access one application while another is running.

The mental model was very useful to the designers while developing the device. It provided them with an operational roadmap on which to base their design decisions. However, its primary value should be to the user. The user should also be able to intuit the mental model underlying the device's design. Once the model is known, the device will be easier to learn and operate. Maybe. That depends on how well the designers created and implemented the mental model.

For the designers' mental model to be of value to the user, it must possess several characteristics:

- It must be visible. That is, the mental model must be suggested by the device's affordances and visibly expressed in the results of the user's actions. These results in turn help the user to intuit the mental model [3]. If the mental model is hidden in the operation of low level design tasks for example, the user will never have an opportunity to learn and employ it.
- It must be based on common, relevant world knowledge. If the mental model is based upon knowledge only the designer is privy to, users will not readily intuit it and it will be of little value to them. For example, the basic desktop metaphor used in the current PC user interface is based on something most people are familiar with: a desk with folders and documents.
- The mental model must be consistently applied to all aspects of the device's operation. All aspects. Period. No exceptions. This is very powerful since it enables the user to extrapolate how to handle newly encountered operations of the device based upon previous experience with the device. Every exception to the mental model creates uncertainty and is a special case the user must remember. This is one of the reasons many people consider the Macintosh computer to be easier to learn and operate than a Windows PC. Windows has many more exceptions in the implementation of its version of the desktop metaphor than does the Macintosh.[19]

[19] We consider one mental model exception when we discuss the Principle of Least Astonishment in the next chapter.

Actions required of the user that are not consistent with the device's mental model will likely be perceived by the user to be confusing and frustrating.

Non-use obtrusiveness: One of the most basic and obvious principles of a transparent use device is that it is transparent when you are not using it. Devices with non-use obtrusiveness violate this most basic principle.

Device non-use obtrusiveness is all about the physical design of the device. In general, the larger, more rigid, or thicker the device, the more non-use obtrusiveness it will have [5].

Non-use obtrusiveness can be especially frustrating since the device is forcing us to take notice of it when we are not even using it for our primary task. It is bad enough when the device forces us to take notice of it when we are using it. It is even more frustrating when we are not using it.

Besides annoying us and making us uncomfortable, non-use obtrusiveness can cost us in real money. The device's non-use obtrusiveness can result in real damage to the device or other things we are wearing such as clothing. When a device catches on a door latch or a seatbelt receptacle, for example, it can be ripped from its place on the person and damage itself and/or that to which it is attached. For example, many times my cell phone has caught on the seat belt receptacle and been pulled out of the holster and fallen to the car floor behind the seat as I was getting out of the car. The problem was that it protruded from my body to a degree and in a manner that allowed the seat belt receptacle to get into the space between it and my body and pry the phone out of the holster.

Incidents like that as well as being poked and pinched by a protruding device or device elements (antenna, etc) can significantly reduce the desire to carry the device. Deciding not to wear or carry an element of a wearable system because of its high level of obtrusiveness can significantly reduce the effectiveness of the wearable system as a whole.

Service Operational Inertia

Although service Operational Inertia is not physical, it is no less frustrating to the user.

Setup Effort: Setup effort for a service or application includes navigating within the user interface of the device to find the application and then launching it. The organization of the menu system has a great impact on this aspect of setup effort. Providing frequently used operations at the top of the menu tree is an obvious strategy to minimize this cause of setup effort.

When using a speech user interface, setup effort includes specifying the spoken command to launch the service. If the interface uses a prompt and response command format, where the system prompts the user for every element of the command, the setup effort can be significant. In a more conversational interface, the main difficulty is remembering the correct structure and elements of the command so it is spoken correctly.[20]

If the service requires authentication of the user, this can contribute significantly to setup effort. Specifying passwords, PINS, and other authentication codes focuses the user's attention on the implementation aspects of the security service rather than the user's primary task (which is almost never to use the service for its own sake).

Providing parameters or other service specific information required to configure the service increases setup effort. This is an example of a support task since the relevance to the user's primary task can be apparent.

Closely related to providing parameters for a service is entering the same information in multiple parts of the service setup or configuration areas. We have probably all experienced at one time or another the frustration of calling customer support, being asked for information and then being

[20] This in itself is part of the Speech User Interface's interaction complexity.

transferred to another area of the support service and being asked for the same information. This is setup effort gone amok. As we are giving the same information for the second or third time we are silently screaming "Can't you people or your computers talk to each other?".

The necessity of entering the same information multiple times is due solely to the implementation of the company's support service (usually computer systems that don't share the same databases). And that is driven mostly by what is best for the company, not the users.

Other elements of setup effort for a service include configuring other devices required by the service and configuring or terminating other services as required by the service. An example of the former is adjusting a microphone before using a speech recognizer to take into account the level of ambient noise. An example of the latter is having to terminate all running programs when installing a new application.

Interaction Complexity: Interaction complexity is the measure of how difficult it is to use the service after it has been properly set up. A major determiner is the complexity of the service's commands and the consistency of the command format.

We discussed the role a consistent application of the mental model has in determining a device's interaction complexity. This is perhaps even more important for a wearable application or service. One of the things unique about a wearable is that the user is usually doing something else when using it. For example, the user may be walking on a busy sidewalk and using the wearable to check on an appointment. His attention must not be focused on the wearable since he must focus on navigating the crowded sidewalk. This *divided attention state* makes any command that does not follow the standard command format much more difficult to use, even more difficult than if you encountered it seated before your PC to which you can give your complete attention.

Navigating long application menu trees or becoming unsure of where you are in the application will also contribute to interaction complexity. Location is context. And if you cannot quickly determine where you are in the application or system, the probability of making an error significantly increases.

Non-use obtrusiveness: For a service or application, non-use obtrusiveness includes the disk space taken up when it is not being executed. Some applications such as development environments require over a gigabyte of disk space when not being used (and even more when being used).

Service and application non-use obtrusiveness also includes any system resources such as memory that are not released after the program terminates execution. Such memory leaks accumulating over time can compromise the performance of other applications and can even cause them or the computer to crash.

System Operational Inertia

In a system composed of multiple components, the Operational Inertia is made up of two parts: the operational inertia of each of the components and that arising from the interaction among the components in the system.

Setup Effort: For a system setup effort involves three parts

1. Putting on each device, the so-called 'gearing up' effort (and its counterpart, 'tearing down', that is, taking off all of the devices). This is perhaps the most serious impediment to distributed wearable systems and the impediment increases with the increasing number of devices making up the distributed system. Besides the actual actions of putting on the devices, system setup effort includes designating the primary I/O devices and configuring global user preferences. This impediment is one of the primary attractions of physically integrated devices.

   Setup effort also includes transferring information from one device to another when you wish to change devices in a system. And it includes removing a device from one system and adding it to another, for example, when you change clothes and move your cell phone from the garment you had on to the one you are putting on.

2. Interconnecting devices as part of the system. This includes physical connections (wires, cables, physical connectors) and wireless connections (authentication of Bluetooth devices).
3. Performing system maintenance. This includes charging the devices or replacing batteries, removing the devices to wash the garment in which they are embedded and replacing them, and authenticating yourself to the system before use.

Researchers have tried different methods to reduce or eliminate the setup effort for wearable systems. One is to host the entire system, or much of it, on a single garment. This garment has been a vest, a shirt, and a jacket. This can be as extensive as the MIThril vest (left in Figure 4-3), or as sparse as the SmartShirt (right in the figure).

Pre-attaching the wearable system to a garment such as a shirt, vest, or jacket solves most of the system setup effort. However, it may make the garment itself more difficult to put on. This would be especially true for close fitting garments such as a shirt. This option is most useful when the elements and configuration of the system do not change significantly while the system is being worn. That way, most of the setup effort for the system is incurred only once.

Clothing presents unique issues with setup effort. The more fine grained the embeddedness is, the harder it is to access and retrieve the devices[21]. Beyond a certain level of emdeddedness, the devices or elements become so difficult to remove and use or maintain outside of the garment that they should be regarded as permanently embedded.

[21] Chapter 3 discussed the different levels of embeddedness for electronics in clothing. This ranges from what we call 'packed' – simply hanging devices on a garment, to "fine grained integrated" where device components are themselves disaggregated and embedded.

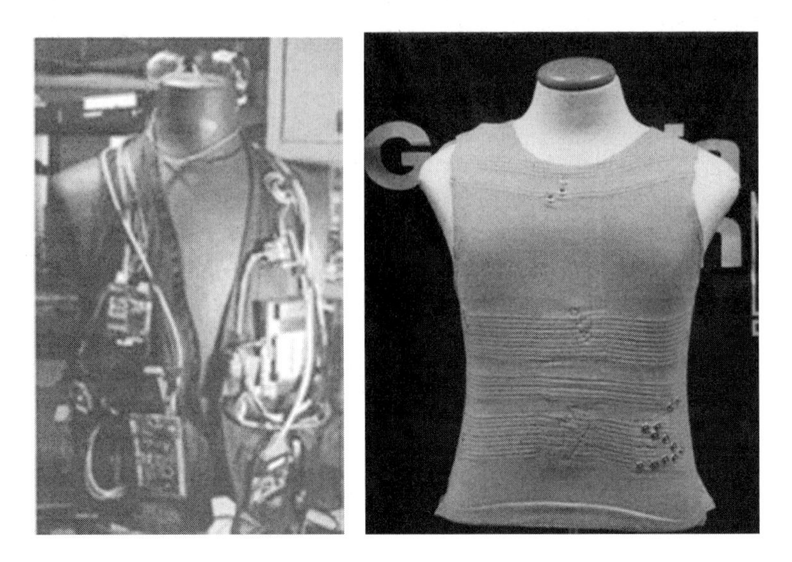

Fig. 4-3. Hosting a Distributed Wearable Systems in a Single Garment. Left: MIThril wearable system *(MIT Media Lab/Alex Pentland)*, right: GATech Smartshirt (*Courtesy: Textile Information Systems Research Laboratory, Georgia Institute of Technology, Atlanta, Georgia, USA)*

Interaction Complexity: Interaction complexity reflects how difficult the system is to use once it is set up. System level interaction complexity is caused by deficiencies in the way the devices of the system interact with one another. One cause is the PAN being slow or the communications protocol being overly complicated and requiring excessive processing by the nodes of the network. The result for the user is poor response by the system to user requests.

One of the most visible elements of a wearable system, and a main contributor to interaction complexity, is the user interface. It is imperative that the system present a single, consistent interface consisting of a common set of interface mechanisms to the user. This means that the user will have to learn only one interface as opposed to a different one for each device. Here the design of the wearable system's central unit is crucial. Ideally, it would specify a common set of user interfaces that all third party system devices would adhere to.

Creating a user interface that is both consistent and intuitive can be especially difficult when dealing with a multi modal interface. Multi-modal interfaces combine multiple interface mechanisms, for example speech,

GUI, and gesture. In the best of these, the user can move from one interface mechanism (say speech), to another (GUI) within the same command sequence.[22] The user can also use multiple mechanisms simultaneously to increase the performance of the interface as a whole as in using a touch screen to point to an item while using speech to identify it. The use of the touch screen can reduce the effects of misrecognition by the speech recognizer [6].

Non-Use Obtrusiveness: When I am not using the system, I expect to not have to think about it.[23] The degree to which the system makes me aware of its presence when I am not using it is the measure of its non-use obtrusiveness. Most non-use obtrusiveness is caused by the individual devices. However, system non-use intrusiveness could be increased by the poor placement of connectors or attachment points on the clothing for the devices, especially if wires are used to connect some of the system elements. This would make the devices obtrusive, regardless of how well they themselves are designed.

System OI is tough to minimize because, in many cases, it does not become apparent until after the system is designed and trialed. This is made more difficult when you have an open system that accepts devices from third parties. This raises the possibility of new types of devices that were never envisioned when the system was first created, increasing the system OI when they are used.

[22] We discuss wearable user interface design in more detail in Chapters 8 and 9.

[23] This is true even if the system itself is operating, performing support and device tasks.

Learning Operational Inertia

Another source of OI is Learning OI. This is the effort to initially learn how to operate the device, or service or how to use elements of a system together. Learning OI can be a powerful disincentive for using a device or system. It is often the reason why people, especially casual users, do not use many of the features of a device or system.

Portable electronics can have significant learning OI. Devices tend to be small, with small screens, making it difficult to provide multimedia and information rich tutorials or instructions on the device itself. Highly integrated devices can pose significant learning impedance due to the sheer number of features and their interactions. Cell phone manuals often have 200+ pages.

Setup effort of Learning OI includes finding and setting up the learning materials. This can include using other devices with their own OI, such as a PC on which to view the learning materials.

The difficulty the user has in using and understanding the learning materials determines its Learning Interaction Complexity. Even if the learning material is easy to use, that is, navigation, selection, etc., is easy, if the material is hard to understand, the learning OI is high.

Except for the space (physical and/or disk) required to store the learning material, there is little non-use obtrusiveness. However, in memory constrained devices, the need to keep a large learning file in memory can be obtrusive since the user may not be able to load other desired applications.

If the device, service, or system interaction complexity is high, the user is more likely to re-experience learning OI since they may have to take refresher courses or refer to the learning material for a specific question when the regular help does not suffice.

4.2.3 The Goal: ZOIDs and NZOIDs

So how do we use Operational Inertia in the design of wearable devices, services, and systems? The goal should be to eliminate all sources of setup effort, interaction complexity, and non-use obtrusiveness in the wearables we develop. We have a name for such devices Zero Operational Inertia Device, or ZOID[24].

ZOIDs can be used almost unconsciously. They contain no setup effort and so can be used instantly. They are transparent in their use and have no interaction complexity. That is, the mapping between the actions of the device or service and those required for the actual task we want to perform are 1 to 1 and are what we expect. Finally, ZOIDs have no non-use obtrusiveness. They never make us uncomfortable or constrain our activities. And they always provide the exact information we need, when we need it, and nothing otherwise. Clearly, ZOIDs would not interpose themselves between the user and their real task in any discernable way.

However, we do not know how to build ZOIDs. We don't even know if they can be built. The situation is made even more difficult when we realize that Setup Effort, Interaction Complexity, and Non-Use Obtrusiveness can be interrelated. Decreasing one can increase another. So we must back off of this ideal a bit and talk about Near Zero Operational Inertia Devices, NZOIDs. NZOIDs represent a grudging compromise and become a more achievable goal. How will we know when we have an NZOID? How little Operational Inertia must a device, service, or system have to get this label?

Many people are already using a device that approaches a NZOID. Consider eyeglasses. They have little set up effort; simply take them out of your pocket, open the arms, and put them on. While you are wearing them, they have very little interaction complexity – you simply look through them. And when you take them off and store them, they rarely get in your way. Another NZOID example is a simple analog wrist watch with an expandable band.

[24] For a device to be a ZOID it must have no device OI and each of its services must have no OI. Also, it must not contribute to any system OI.

Note that ZOIDs (and NZOIDs) will most likely be very complex. It is usually very difficult to design something that is very simple to use. However, the result is devices, services, and systems with near zero visible complexity.

4.2.4 Reducing Operational Inertia

Now that we know what Operational Inertia is and have seen some examples, how do we minimize it? The short answer is: with great effort. The long answer is more involved.

Minimizing OI is not straightforward. An obvious reason is that we don't think like this. The conscious, systematic process of identifying and reducing or eliminating setup effort, interaction complexity, and non-use obtrusiveness in devices, services and systems is not something with which we have much experience.

As an example of just how off the radar screen this is, consider a review of a Samsung digital watch with an embedded GPS receiver [7]. A portion of the review summary is shown in Figure 4-4. If you look at the review criteria you will notice a peculiar omission: Ease of use. None of the other criteria, even taken together, cover ease of use. This illustrates the degree to which OI considerations are not even thought of.[25]

Another reason why it is difficult to reduce OI is that there is tension between the OI elements setup effort, interaction complexity, and non-use obtrusiveness. It is often difficult to reduce one without increasing one or both of the others.

[25] To be completely accurate, the reviewer does briefly discuss the weight of the device and how it feels on the wrist which would correspond to elements of non-use obtrusiveness.

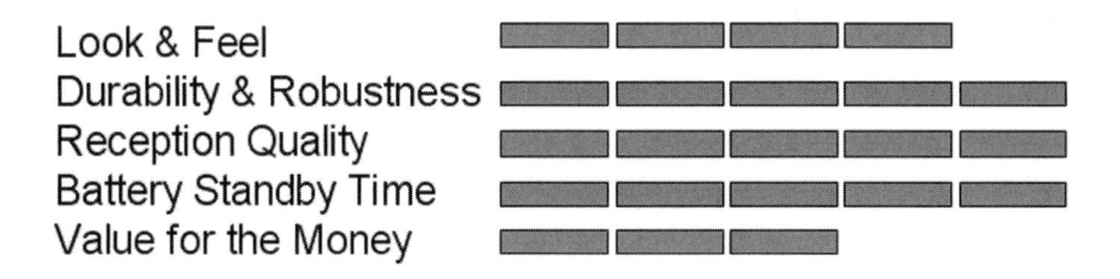

Fig. 4-4. Samsung GSM Watch Review Summary *(from [7])*

Fig. 4-5. Casio Calculator Watch *(courtesy of CASIO)*

Figure 4-5 shows a calculator watch from Casio. The design of this watch greatly reduces the obtrusiveness of a watch and calculator as separate objects. However, doing so has increased interaction complexity of the watch. Because the calculator buttons are so small, the user must focus more attention on pushing the correct one, perhaps even having to use a stylus or the tip of a pen or pencil. This takes their attention away from their real task which required the use of the number the user is calculating with the watch.

As another example, we can often reduce interaction complexity by setting additional user preference or application settings. However, this increases setup effort.

4.3 SYSTEM DESIGN PRINCIPLES

The discussion in the previous section lays the foundation for the definition of a set of mainstream wearable system design principles. These principles are meant to guide the design of devices, services, and the wearable system as a whole.

The principles are shown in Figure 4-6 and are based on the concept of Operational Inertia introduced earlier. The goal of these principles is to promote the design of systems with Near Zero Operational Inertia for the devices themselves, the services the devices contain, and the system as a whole.

4.3.1 Origin of the Design Principles

These design principles are primarily based upon the Person Integrated Communications (PIC) project in the 1997 – 2002 and the follow on Conformables project from 2002 – 2005. The PIC project was focused on understanding how to create devices that remove the perception of distance and separation from the person or device with which they are interacting caused by having to manipulate these devices and services. As a result, communications with people and devices separated by distance would become as easy as if we were standing right next to them.

The concept of Operational Inertia was an outgrowth of the PIC project's objective of minimizing device manipulation. The Operational Inertia concept was refined during the Conformables project. A Conformable is a wearable that has very little OI. It conforms to the user's actions, motion, and needs to such an extent that the user is barely aware of its presence. An

Transparent Use Design Principles

Setup Effort

- Objects incorporating computing & communications retain their basic concept
- Overload normal operational affordances and activities
- Maintain embedded computing and communication functions within the normal usage pattern of their host
- Minimize User Intervention in Setup Processes

Interaction Complexity

- Minimize required user focus on non primary tasks
- Minimize visible complexity
- Respect the inherent I/O limitations of the device
- When incorrect/incomplete information is entered in a fully constrained situation and the required information is known to the system, complete it for the user
- Build Upon / Amplify Normal User Actions
- Design Command Specifications That Are Consistent and Mutually Reinforcing Across All Input Devices of the Same Modality
- Exploit analogies over detailed descriptions
- Minimize POLA violations
- Disaggregate and Simplify

Non-Use Obtrusiveness

- Use body conforming shapes
- Conform to body's contours and motion in all planes
- Do not impede normal operation of body limbs
- Reduce opportunities for conflict with the physical environment
- Attach to the body or clothing in a non-invasive, non-destructive manner
- Maximize Output Information Density

Fig. 4-6. Transparent Use Design Principles

important activity in the Conformables project (and in PIC as well) was observational studies of people in everyday tasks using communications and handheld computing devices. These studies formed the foundation of the design principles.

The principles were developed with several assumptions in mind:

- The device (or service, or system) belongs to a single person, its user. It is considered to be a personal, non-shared item.
- The device (or service, or system) is with person for significant period of time each day
- The device (or service, or system) is used to aid the user in their everyday tasks
- The device (or service, or system) is not the focus nor the end goal of the user's tasks

These assumptions mean that some design elements commonly considered are not emphasized. For example, in Universal Design one of the major goals is to make the device "... usable by all people, to the greatest extent possible, without the need for adaptation or specialized design" [8].

Instead, since the device belongs to a single person, we focus on allowing the user to extensively customize the objects to fit their individual preferences. This is part of Setup Effort and something we aim to make as easy as possible, recognizing that the amount of customization could be significant.

These guidelines address four of the five factors affecting acceptance of wearables (discussed in Chapter 1)[26]:

- Wearability

[26] The fifth factor, price, is not addressed

- Functionality
- Operation
- Aesthetics

The principles also apply to the design of the software architecture, user interface, system architecture, and the system's interactions with the external environment.

The goal of these guidelines is to design NZOIDs (Near Zero Operational Inertia Devices, services, and systems). The guidelines apply to each device in isolation, each service local to a device, the system's inter-element interactions, and to the system taken as a whole.

Relationship with Other Design Guidelines

The transparent use principles focus on how the user performs their daily tasks and how to use the actions involved in those tasks to make the use of wearable devices transparent. This orientation reflects the influence of Activity-Centered Design [9].

Activity-Centered Design (ACD) places activities, not user's characteristics, at center of focus. In this way it differs from Human-Centered Design[27] [10]. In ACD activities are coordinated, integrated sets of tasks. Tasks are composed of actions.

Transparent use design principles focus on a series of tasks performed relative to a goal. The user's primary task is directly related to the goal. But the transparent use guidelines go further and also focus on the user's structure, posture, etc. during the tasks. The guidelines break each task into the three elements of Operational Inertia: Setup Effort, Interaction

[27] Also known as User Centered Design

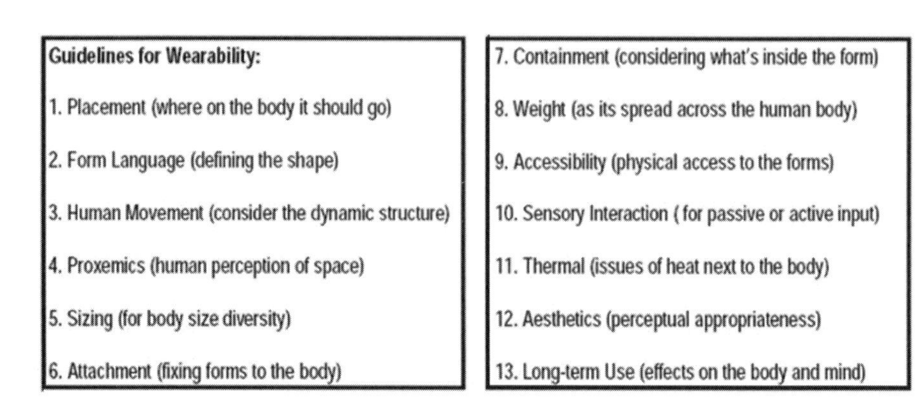
Guidelines for Wearability:

1. Placement (where on the body it should go)
2. Form Language (defining the shape)
3. Human Movement (consider the dynamic structure)
4. Proxemics (human perception of space)
5. Sizing (for body size diversity)
6. Attachment (fixing forms to the body)
7. Containment (considering what's inside the form)
8. Weight (as its spread across the human body)
9. Accessibility (physical access to the forms)
10. Sensory Interaction (for passive or active input)
11. Thermal (issues of heat next to the body)
12. Aesthetics (perceptual appropriateness)
13. Long-term Use (effects on the body and mind)

Fig. 4-7. CMU Wearability Guidelines (from [5])

Complexity, and Non-use Obtrusiveness. These elements reflect usage modes of an entity over time within a task.

Another influence was the wearability study done by CMU wearables group [5]. In that study, the authors looked at the different locations on the body where wearable devices could be worn. They identified those spaces on the human body where solid and flexible forms could be worn without interfering with the user's movement. The report defined a set of guidelines for designing wearable device form factors. The guidelines are shown in Figure 4-7 and are discussed below.

Guideline 1: The report recommends that wearable devices be placed in areas that have a large surface area and are relatively constant in size within the adult user population. These areas should not move very much during the course of the user's day, minimizing the chance that the device will

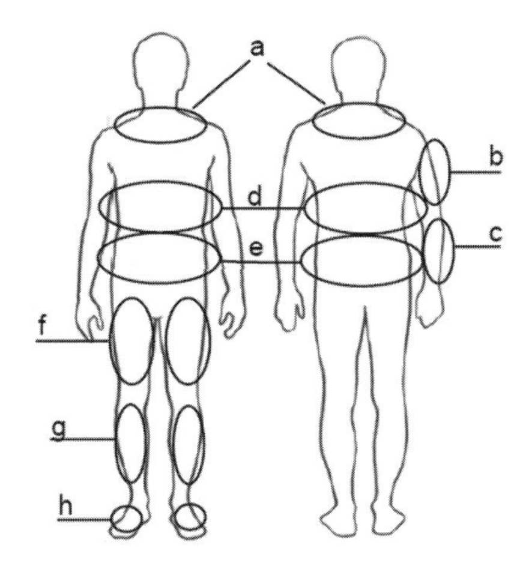

Fig. 4-8. The general areas found to be most unobtrusive for the wearable objects are: (a) collar/necklace area, (b) upper arm, (c) forearm, (d) rear, side, and front of the ribcage area, (e) waist and hips, (f) thigh, (g) shin, and (h) top of the foot, with the addition of the head band area (not shown) specifically from just above the ears to the rear base of the skull (from [5]) *(CMU Wearables Group)*

cause the user discomfort. Based on these criteria, the authors identified the eight areas on the body shown in Figure 4-8 [5][28].

Guideline 2: The shape of the device should have a concave inner surface fitted to the body's contour in the area of attachment. The outsides should be convex so most objects it comes into contact with glance off. All corners should be rounded for comfort and sides should be tapered toward the back to increase stability on the body.

[28] Interestingly the wrist, considered by many to be the most valuable piece of real estate on the body for wearables, was not included in the list of areas.

Guideline 3: In the course of a day the body can assume many different positions. These can involve bending, twisting, and rotation among the various areas of the body such as the torso, waist, legs, and arms. A wearable device must accommodate these movements without causing undue discomfort to the person.

Guideline 4: The brain accounts for an envelope of space surrounding the body. Depending on the area of the body this envelope can extend less than 1 inch to about five inches from the body. Anything within this envelope is perceived to be part of the body [11].

Guideline 5: People come in different sizes and shapes. Indeed, a person's own size and shape can vary noticeably over a period of time ranging from a day or longer. And, as a person ages, their size and shape can change significantly. A wearable device should be adjustable so it fits the current shape of the body as closely and comfortably as possible. Segmented and articulated form factors can provide some accommodation.

Guideline 6: The method used to attach a device to the body can also greatly influence the device's perceived comfort. Single points of attachment are unstable and allow the device to shift and rotate in ways that can become obstructive. If a device follows the body's contour throughout its length and width, it will be much more stable and more likely to move in concert with the body's own movements.

Guideline 7: Any device will contain components such as pc boards, batteries, switches, etc. These components are rigid and can make it difficult for the device containing them to accommodate user movement through flexion. The emerging use of flexible pc boards and component mounting technology will reduce this problem. In addition, segmenting the electronics into smaller units connected with flex cable supports segmented, articulated designs.

Guideline 8: The weight of the device must be considered when deciding where to place the device on the body. The heavier the device, the closer to the centerline of the body it should be placed. Placing heavy objects far out on a body limb such as the wrist or ankle can rapidly result in muscle fatigue. Heavy objects should be placed on the waist and hips as these are the best load bearing areas of the body.

Guideline 9: Accessing the device must be intuitive and not require excessive user focus. Wearables are used in the performance of the user's primary task. Using the device itself is rarely the primary task. Therefore, the user will want to concentrate most attention on the primary task, be it making dinner, going somewhere via walking or driving, etc. This state of divided attention makes easy and intuitive access to the device and its functions critical to its usefulness.

Guideline 10: Sensory interaction is the domain of the user interface. Multiple IO modalities – speech, visual, audio, tactile, etc should be available and the device should choose whichever will be most effective given the current user context. The user should be able to switch from one modality to another at any time upon request.

Guideline 11: Thermal issues with a wearable will become more important as wearable devices are worn for longer times. The close proximity with the body, especially if they follow the body's contour along their length and width as recommended above, decreases the opportunity for air flow across the surface of attachment. This can interfere with cooling. Long, sustained periods of contact with a device at high temperature can cause problems.

Guideline 12: Aesthetics of a wearable device will be quite important if the device is visible. The device should not violate the user's sense of fashion. This can be a moving target depending on where and when the device is worn. The look required for a formal restaurant is quite different from that suitable for casual time at home.

Guideline 13: The long term effects of wearables are unknown. However, the recent experience of Steve Mann at the Toronto airport is revealing [12]. Mann has worn his wearable system almost all day and nearly every day for over twenty years. When his system was suddenly removed from him by airport security guards, he experienced disorientation. Walking to the gate without the use of his system, he claims to have fallen twice and passed out once. He boarded the plane in a wheelchair. This may represent an extreme case, but it does indicate that the long term effects of using wearables are something that must be studied.

Many of the CMU Wearability guidelines map into the transparent use design principles. This mapping is shown in Table 4-1. For example, the wearability guideline 'Form Language' maps directly into the transparency use design principle 'Use body conforming shapes". Indeed, the CMU guideline was the main contributor to that design principle.

Similarly, the wearability guideline 'Proxemics' maps to the transparency design principle 'Do not impede the normal operations of the body's limbs". The guideline "Human Movement" maps into "Conform to the body's contours and motion in all planes".

Table 4-1. Guideline Mappings

CMU Wearability Guideline	**Transparent Design Principle**
1. Placement	*No direct mapping*
2. Form Language	Use body conforming shapes Reduce opportunities for conflict with the physical environment
3. Human Movement	Conform to the body's contour and motion in all planes
4. Proxemics	Do not impede normal operation of body limbs
5. Sizing	Adapt to body's contour and motion in all planes
6. Attachment	Use Body Conforming Shapes Attach to body or clothing in a non-invasive, non-destructive manner
7. Containment	Conform to the body's contour and motion in all planes

	Use body conforming shapes
8. Weight	Do not impede normal operation of body limbs
9. Accessibility	Minimize user's cognitive load Minimize user focus
10. Sensory Interaction	Minimize user's cognitive load Minimize user focus
11. Thermal	*No direct mapping*
12. Aesthetics	*No direct mapping*
13. Long Term Use	*No direct mapping*

For some of the CMU wearability guidelines there is no direct mapping to a transparent use design principle. This is because either the wearability guideline is subsumed in multiple transparency design principles or because the guideline is simply not addressed by the principles. An example of the former is the wearability guideline 'Containment'. This is subsumed by the transparency design principles 'Conform to the body's contours and motion in all planes' and 'Use body conforming shapes'. As an example of the latter, the wearability guideline 'Long Term Use' is simply not addressed by the principles.

In addition, there are many transparent use design principles that are not addressed by the wearability guidelines. This is because the principles address more than just wearability. They also address minimizing the effort required to setup and maintain the device, and minimizing the effort in interacting with the device. Finally, the principles can be applied to services and applications, which have no physical manifestation.

In the next chapter we consider the principles and their application in detail.

REFERENCES

[1] Norman, Donald, A., 1998a, The Invisible Computer, MIT Press, Cambridge, MA, p75.

[2] Norman, D., 1989, the Design of Everyday Things, Currency Doubleday, New York, pp 50-52.

[3] Norman D., 1998, p196.

[4] Norman, D., 1989, p9.

[5] Gemperle, et. al., 1998, Design for wearability, Second International Symposium on Wearable Computers, 19-20 October 1998, Pittsburgh, Pennsylvania, USA, pp. 116-122.

[6] Ehlert, P., 2003, Intelligent user interfaces, Research Report DKS03-01 / ICE 01,, Mediamatics / Data and Knowledge Systems group, Department of Information Technology and Systems, Delft University of Technology, The Netherlands, pp 20, 21.

[7] Ciao Shopping Intelligence, 2004, Brava two zero this is oscar two zero come in, over, http://www.ciao.co.uk/Samsung_GPRS_Watch_Phone__Review_5422565, accessed March 23, 2006.

[8] Rose, Bettye, et. al., The Principles of Universal Design, Version 2.0 - 4/1/97, North Carolina University, http://www.design.ncsu.edu:8120/cud/univ_design/principles/udprinciples.htm, accessed 2/23/06.

[9] Norman, Donald, Human-Centered Design Considered Harmful, 2004, http://jnd.org/dn.mss/human-centered.html, accessed 2/21/06.

[10] Usability Professionals' Association, What is User-Centered Design? http://www.usabilityprofessionals.org/usability_resources/about_usability/what_is_ucd.html , accessed 2/23/06.

[11] Hall, E. T., (1990), The Hidden Dimension, Anchor Books.

[12] Guernsey, Lisa, At Airport Gate, a Cyborg Unplugged, New York Times online, http://tech2.nytimes.com/mem/technology/techreview.html?res=940CE0D71239F937A25750C0A9649C8B63, March 14, 2002, accessed February 21, 2006.

Chapter 5

MAINSTREAM WEARABLE DESIGN IN DETAIL

5.1 THE TRANSPARENT USE DESIGN PRINCIPLES IN DETAIL

We now discuss the transparent use design principles in detail. We give examples of their application and note their limitations. It is important to realize that these principles should be used in concert with others. The principles do not cover every aspect of device and service design nor are they necessarily complete in addressing their subjects.

It is also important to realize that these design principles are very much a work in progress. We have made modifications in the principles and will continue to do so as we gain more experience in their application. The principles are grouped according to the components of Operational Inertia: Setup Effort, Interaction Complexity, and Non-use obtrusiveness.

5.1.1 Setup Effort

The setup effort guidelines are aimed at minimizing the effort required to get a device, service, or system ready for its intended use, and to minimize its maintenance.

Objects Incorporating Computing & Communications Retain Their Basic Concept

Utilizing embedded electronics in an object should not require that the user treat it in a way that either violates or calls into question the basic concept of the object without the embedded electronics.

For example, using the embedded electronics in jewelry (the necklace in Figure 5-1 with electronics in the front and the battery in the back) should not require the user to treat it differently from a regular necklace.

As another example, consider the ICD+ jacket discussed in Chapter 3. This design principle dictates that the user treat it as he would any jacket without embedded electronic devices even when using those embedded devices. The maintenance of the electronics cannot require operations that would contradict the host object's essential properties – those properties that make the object what it is. For example, the ICD+ jacket includes embedded electronics and I have to opcn up thc lincr and rcmovc all of the electronic components before I wash the jacket. By doing this the basic concept of the jacket has been compromised. The setup effort to maintain the jacket (removing all of the electronic components) has focused the concept on the jacket as a host for the electronics rather than a garment. I should also not have to 'plug in' the jacket to recharge the batteries, nor have to go through the extra, explicit steps to authenticate myself to the jacket's controller to access its data.

But how can we find transparent ways to accomplish those actions required by the embedded electronics of the jacket? If we are not going to remove the electronics for washing and drying, we must do several things:

- Enclose the electronics and wiring to protect them from the water and chemicals used in washing
- Provide efficient thermal protection and heat transfer away from the electronics so that they are not damaged during ironing or hot water cycles of the wash.
- Provide multiple paths between components. Washing and ironing will put a significant strain on the physical integrity of the boards, components, wires, and couplings of the devices. It is inevitable that over the course of its life one or more of those interconnections will break. In

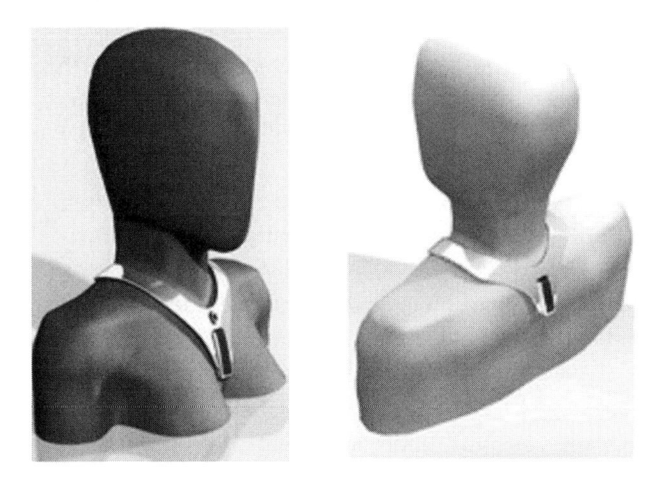

Fig. 5-1. An Electronics Enhanced Necklace *(Motorola Inc.)*

that case, the path between the components must be rerouted to another path that is still intact.

One thing becomes clear: for the user to be spared the need to focus on the embedded electronics, the electronics must be permanently embedded and never need to be accessed or removed from the jacket.[29] This has implications on the design and the type of electronics that will go into jackets. We discussed this in detail in Chapter 3 when we looked at the issues of embedding electronics in garments.

[29] However, we need not design against catastrophic failure; for example, the failure of a major system component or circuit board. In case of such a failure the jacket would go to an 'eTailor' for repair, much like we would take an unenhanced jacket to a regular tailor in the event of a major split or tear in the jacket's fabric. Indeed, it is possible that tailors in the future will provide 'eTailoring' service and have an eGarment technician on staff.

The device can exhibit capabilities based on the embedded electronics as long as those capabilities are complementary with or orthogonal to the device's unenhanced capabilities, in which case the concept of the host device (ex. a jacket) would formally expand to include the new capabilities.

For example, a jacket can have a control panel sown into a sleeve that controls the various embedded devices without compromising the jacket's mental model. The panel adds new capabilities to the jacket's capabilities involving manipulations of the sleeve. Since the unenhanced jacket's mental mode does not contain any sleeve manipulations, the new capabilities extend but do not violate the jacket's mental model. Contrast that with having to manipulate the embedded electronics before washing the jacket. This violates the jacket's mental model since typically nothing is removed from the unenhanced jacket before washing it.

But if we are not going to be able to access the components and don't want to even focus on them, then how do we maintain the electronics, for example, recharge the batteries or authenticate ourselves to the controller to gain access to the information it contains? This is where the next two design principles come in.

Overload Normal Object Operational Affordances and Activities

Every device has unenhanced physical or logical affordances that can be used to maintain or interface with its electronic components. By utilizing these elements in their normal usage pattern, we can interact with the device's electronics in a nearly transparent manner.

For example, consider the jacket in the previous example. It contains a device that has personal information such as biometric data acquired from body worn sensors and a record of my movements provided by location based services (GPS, location beacons, etc.). I do not want anyone other than myself to get access to this without my permission. This requires I authenticate myself when I put on the jacket. However, since authentication of my identity is not something I associate with a jacket, this should be transparent to me.

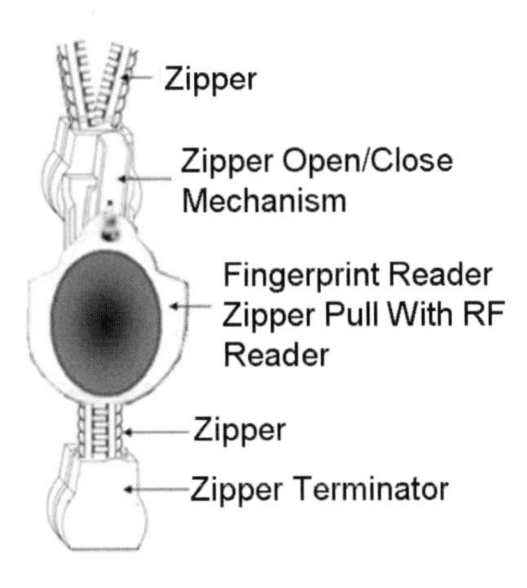

Fig. 5-2. Authenticating Zipper Pull *(Motorola Inc.)*

One possible approach, shown in Figure 5-2, is to overload the function of the jacket's zipper pull. Integrating a small fingerprint reader in the pull and providing a signal path along the zipper allows the jacket to read and authenticate my identity by reading my fingerprint while I am zipping up the coat. By overloading the typical position of my thumb on the zipper pull when I use a zipper, I can provide the information required for authentication almost transparently while I zip up the jacket.[30]

There are other ways to make authentication almost transparent. Using speaker identification (SI) allows me to authenticate myself via my voice. However, to make it transparent, we must go beyond the capabilities of current SI systems.

[30] The requirement to actually zip up the jacket is itself a limitation of this technique. If the weather is warm and I want to simply put the jacket on without zipping it up, the authentication becomes non-transparent since I must place my finger on the zipper pull to authenticate myself even though I am not zipping the jacket up. So we have not yet made the process completely transparent.

Current SI systems require a user to repeat back a randomly chosen phrase known to the system. The phrase is randomly chosen so that someone else can't spoof the system with a tape recording of your voice saying a phrase that never changes.

However, making the user speak a word or phrase that has nothing to do with their primary task increases the setup effort. Instead, the system should allow the user to simply give a command for one of the tasks she wishes to do and use that speech to identify the user. By piggybacking on a command, the authentication via SI is transparent[31].

Another example is the work on dual purpose speech described in [1], [2]. With dual purpose speech, the user continues to engage in a conversation with another person while his wearable monitors his speech. The user's conversation with the other person contains sufficient information to allow the wearable to take action on the behalf of the user for tasks within the current context. The example the authors give is the user being in a conversation with someone and discussing when to meet again.

As the two people discuss when to have the next meeting the user's wearable monitors the conversation and activates the calendar application. By the time the two people decide on a date to meet again the use's wearable has obtained sufficient information from the conversation to insert the meeting in the user's calendar.

The key to the calendar application's transparency is that the grammar driving the speech recognizer for the application understands the speech the user employs as part of the natural conversation with the other person about

[31] There may be places where this mechanism can't be or should not be used. If I am putting on my coat in the library, I should use a different mechanism than speech to avoid disturbing others. The need to consciously choose an alternate authentication mechanism introduces a small amount of non-transparency, regardless how transparent the alternate mechanism is.

when to meet again. Thus the user's speech is overloaded to create the reminder in the wearable for the next meeting. The user never explicitly focuses on the calendar application[32].

Maintain Embedded Computing And Communication Functions Within The Normal Usage Patterns Of Their Host

Maintenance falls under setup effort since it is a periodic activity that must be done to ensure continued use of a device or system. Maintenance activities also have little to do with what we really want to do with the device or system and so we need to find a way to make them as transparent to the user as possible.

As a case in point, we still have not dealt with the problem of transparently charging the jacket's batteries discussed above. The best approach is to charge the batteries as a side effect of using the object as its unenhanced concept dictates. To do this we use the actions we naturally use with the unenhanced object to maintain its embedded devices. For example, kinetic watches contain internal electrical generators and use wrist motion while worn to generate power [3]. Another example is generating power from piezoelectric circuits in shoes while walking [4].

In the case of the jacket, we can charge its batteries when hanging it up. This involves a redesigned hanger and power delivery system. Figure 5-3 shows an early prototype of the hanger [5] and a potential refinement [6]. In the refinement power is delivered to the hanger via a special current carrying closet rod or hanger peg. The hanger fits into a channel along the length of the rod to make contact with the power lines, minimizing the shock hazard.

[32] The transparent implementation of dual purpose speech is not yet practical due to the insufficient accuracy of today's speech recognizers. However, it is an attractive approach for the future.

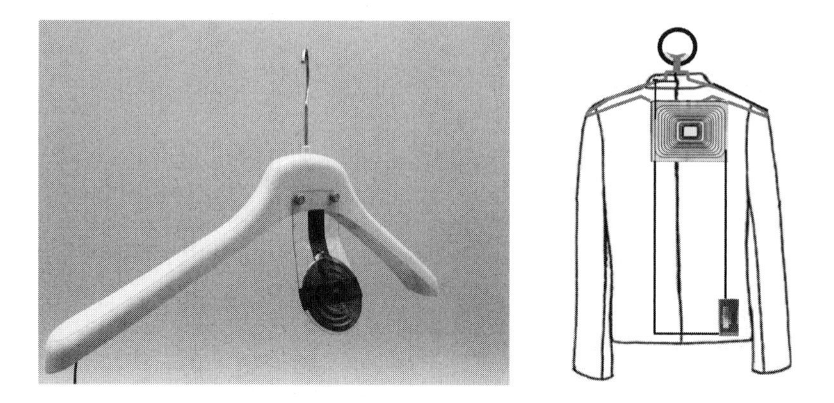

Fig. 5-3. A Power Hanger System. Left from [5] (*Tempere University of Technology, Kankaanpää Research Unit for Wearable Technology, Jämintie 14 FI-38700 Kankaanpää, Finland*) right: a refinement with garment *(Motorola Inc.)*

The hanger contains a micro switch that closes under the weight of the jacket, completing the circuit and allowing power to flow from the rod to the inductive plate on the hanger. The time varying current through the hanger's inductive plate induces a current in the jacket's plate which is transformed into the charging current for the batteries.

This overloads the action typically associated with a regular jacket – hanging it up – to charge the batteries. There is nothing to plug in, attach, or connect. This makes the charging task almost transparent. Issues that can work against the transparency are the time it takes for the inductive charging to charge the batteries, procuring the special hanger and rod system, and potential difficulty draping the jacket over the hanger due to miss-alignment between the hanger and jacket inductive plates.

Minimize User Intervention in Setup Processes

Many services performed by a mainstream wearable system will be in the background, for example, monitoring the user's environment, or connecting with new devices that come within the range of the system's short range wireless network. Most of these tasks are not primary to the user.

That is, they are not directly associated with what the user really wants to do; they merely enable the primary tasks. Since they are not primary tasks, the user really does not want to deal with them. Therefore the system should either find alternate mechanisms for performing them or obtain the information required to perform them without involving the user's input.

Using information about the user's environmental and situational context as well as the user's preferences can reduce the amount of information the user must supply and can minimize the number of times the user must be involved in setting up or modifying a service. Using rule bases and reasoning algorithms enables the system to determine what type of information is useful to the user at any point in time. IIowever, this proactive, semi autonomous operation requires that the system provide on demand visibility to and explanations for the actions it took and the decisions it made.

This is especially true of the short range wireless network. Within a pervasive computing environment, a wearable system will encounter many devices that can join its network. Many of these devices will be useful to what the user is currently doing. Automatic authentication for trusted devices makes ad hoc networking much less obtrusive. The system should not ask the user every time it wishes to access one of these devices. Rather, the system must maintain rule bases and algorithms to allow it to determine what type of information is useful to the user at any point in time and automatically authenticate and communicate with the devices within the pervasive computing environment providing that information.

In addition to increasing setup effort, constant requests for user intervention increase the cognitive load on the user. This increases non-use obtrusiveness when the device is not being used. When the device is being used, it increases interaction complexity. We now discuss mechanisms to minimize that.

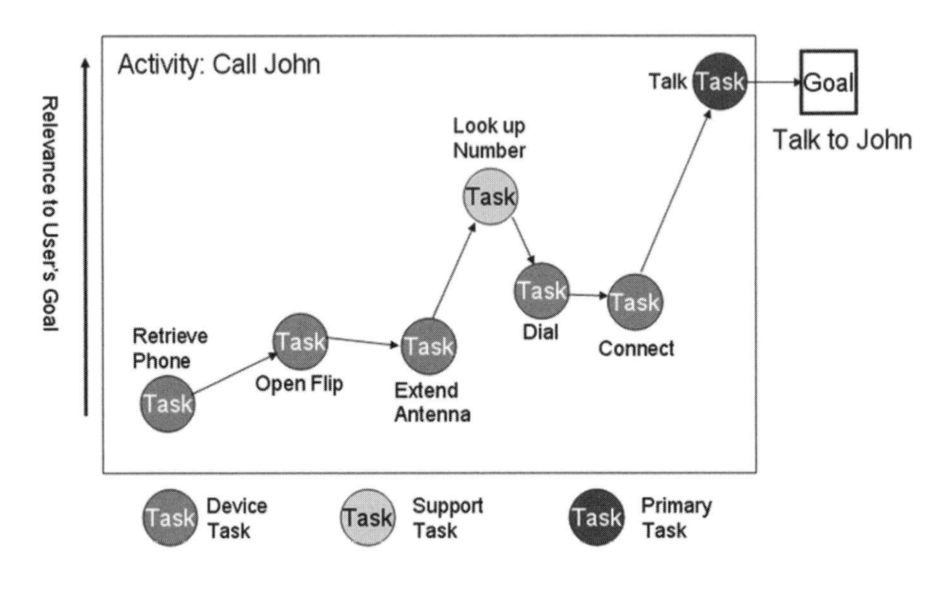

Fig. 5-4. Activity Task Analysis

5.1.2 Interaction Complexity

The goal of transparent use requires that the interaction complexity of a device, service, or of the system be extremely low. The objective is to minimize how much attention and mental effort the user must apply when using the services or devices of a wearable system. The more attention the user must apply to how the task is to be done, the greater the Interaction Complexity.

Minimize Required User Focus on Non - Primary Tasks

The more the user must focus on using a device, service, or system, the greater the cognitive load it imposes on the user. If the user must think about how to use the device, service, or system, they cannot adequately address their current, primary task.

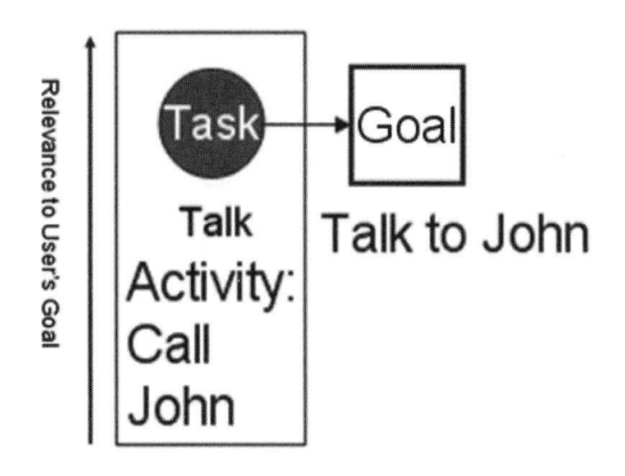

Fig. 5-5. Activity With Only Primary Tasks

As discussed in Chapter 4, an activity is typically composed of tasks that are directly relevant to the goal (primary tasks), tasks that are in support of the primary tasks, and tasks that must be performed solely due to how the device is designed (device tasks). What we should strive for is to make sure that each activity only contains those tasks that are directly related to the user's goal for the activity, i.e. primary tasks.

Figure 5-4 shows the activity 'Call John' which is directly related to the user's goal of 'Talk to John". The activity is composed of the tasks shown in the activity's box. Some or all of them are part of the effort required to use a cell phone for the activity.

Most of the tasks are device tasks - tasks the user must do solely due to the design of the device or the system in which it operates. They have nothing to do with the user's goal. Only one task, 'Talk', is directly related to the user's goal and is the primary task for this activity. If we strip away all of the non-primary tasks so the activity contains only those tasks directly related to the user's goal, we have the activity shown in Figure 5-5. Here, the cognitive load of the activity is determined solely by the cognitive load of the primary task.

There are three elements that make up the cognitive load the user experiences (adapted from [7]). Intrinsic cognitive load is the mental effort the user must expend due to the inherent complexity of the primary task. For example, driving has a higher intrinsic cognitive load than sitting and watching a comedy on TV. The intrinsic cognitive load cannot be reduced by the design of the wearable system or by eliminating external distractions.

Germane cognitive load is the additional mental effort the user must expend in performing the primary task due to the way the wearable system presents information in support of the primary task. Examples of germane cognitive load include how concise and clear information is displayed in visual interfaces, the ease of navigating through the information, and the effort required remembering commands. This is completely determined by the design of the wearable system's user interface mechanisms and I/O affordances.

Extraneous cognitive load for a wearable system is the mental effort the user experiences in performing the primary task due to those factors that have nothing to do with the activity[33]. This is additional mental effort the user must expend in performing the activity to overcome the distracting influences of the environment or the wearable system itself. Examples include the ambient noise, walking while using the wearable system, and being aware of the discomfort caused by the excessive weight or poor placement of elements of the wearable system.

Whereas intrinsic and germane cognitive load are related to performing the primary task, extraneous cognitive load has nothing to do with it. Some causes of it are related to the non functional aspects of the wearable design, such as system obtrusiveness (both in use and non-use). Other causes can be

[33] The definition of germane and extraneous cognitive load presented here differs slightly from those presented in [7]. They were concerned with instructional materials. The mobile nature of a wearable system requires the slight changes we have adopted here.

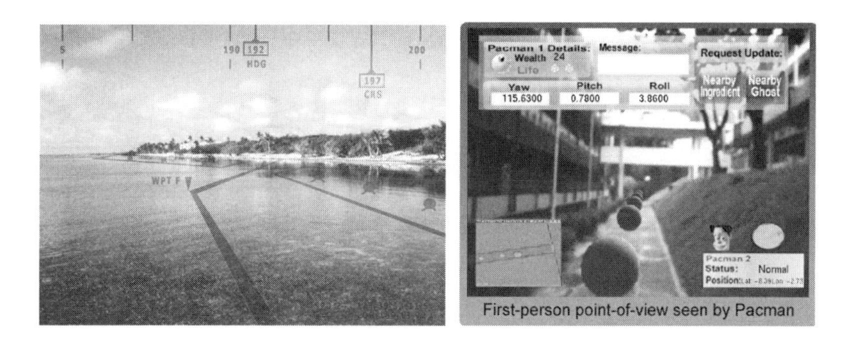

Fig. 5-6. Left: A Sparse Augmented Reality (AR) View ***(photo courtesy of Information In Place)***, and, right, a dense AR View *(Mixed Reality Lab, National University of Singapore)*

dealt with by changing or moving to a new environment in which the primary task is performed.

Minimizing cognitive load is especially important for a wearable system because there is often limited visual context when performing tasks. The user is typically mobile and often does not have a display in front of them to provide a persistent record of the actions and the context in which they occur. In addition, most wearable services are used in support of the user's primary task. They are usually not the main focus of the user's attention. Thus the cognitive load must be less than if the user's full attention was directed at the task.

This requires presenting information in ways that can be rapidly assimilated by the user. For example, use large fonts and small groups of text on heads up displays. For most cases, immersive displays will require most of the user's attention and will hinder the user's ability to use the information to support the primary task. Graphical IO must be understandable 'at a glance'.

As another example, when using Augmented Reality, care must be taken so that the overlaid information does not overly obscure the user's field of view (Figure 5-6). The more information overlaid, the larger the cognitive load is put on the user to assimilate the information along with the external

Fig. 5-7. Using Magnets for Alignment *(Motorola, Inc)*

view. While the denser information overlay of the figure on the right is entirely appropriate for games like Human Pacman [8], it would be problematic for everyday tasks.

Build Upon / Amplify Normal User Actions

Amplifying or aligning the user's actions for the primary task can reduce the focus on using the wearable device or service. For example, magnets in the charger base in Figure 5-7 interact with the opposite polarity magnets on the device bottom to guide the device onto the charging pads, reducing the user focus required to align the device and charger. The magnets amplify and align the user's action of placing the device on the charger. If the device approaches the charger in the wrong orientation, the approaching magnets on the device and charger will be of the same polarity and the user will feel a slight repulsion. Upon feeling this subtle cue, the user simply reverses the orientation of the device.

Another example of Build/Amplify is Apple Computer's recently introduced the MagSafe Power Adapter which uses magnets to secure the power connector to the MacBook's power port [9]. As the user brings the power plug close to the MacBook's power port, the magnets capture and align the adapter plug, ensuring a secure connection.

When Incomplete/Incorrect Data Is Entered and Required Information Is Known To the System, Complete It for User

Sometimes the user enters incorrect or incomplete information. If there is only a single choice in the information that is applicable and required to complete the task correctly and the system knows this information, the system should enter it to allow the user to complete the task.

An example of this is a geographical area with a dialing number plan overlay. In such areas, part of the area code corresponds to a long distance numbering plan where a '1' and area code must be entered. In other areas, no '1' and area code are required[34]. If the user calls a seven digit number covered under the long distance overlay (the number can correspond to the house across the street) a recording comes on telling him that he must first dial a '1' and the area code when dialing this telephone number. This is the only correct course of action. Since the switch already knows the information required to allow the task to be completed, it should simply add the '1' and area code to the seven digits dialed[35]. However, currently, after the message is played, the user must hang up and follow the message's instructions.

It can be argued that the person must be told so that he can learn the rules and enter the correct information upon subsequent issuances of the command. However, in many cases, the user will be issuing the command infrequently or there are many instances of the situation and thus he is likely to make the same mistake each time. When the addition/correction of the command has no effect other than to allow the command to be executed, it is more transparent for the system to simply complete the command with the

[34] There are other variations; for example, the area code, but not a '1' is required

[35] Wireless networks already do this in some situations. For example, if you are making a local call using your cell phone and dial a '1' and the area code, the switch ignores it and completes the call with the remaining seven digits.

correct information. If desired, the system can tell the user how it is correcting the input to allow the command to be executed. The completion/correction of the command can be logged for later viewing by the user, at which time he can decide if it is worth expending the effort to learn the correct command.

We tend to do this ourselves. When someone tells us something and some details are obviously incorrect and we know what the correct information is, we go ahead and use the correct information. We don't correct the other person, since we are sure that they really meant to give the correct information which we know.

Minimize Visible Complexity

Every object has an inherent complexity which cannot be reduced. This inherent complexity is the cause of inherent cognitive load. The goal is to minimize the complexity experienced by the user, regardless of the object's inherent complexity. This visible complexity is the cause of germane cognitive load.

Most mainstream wearable devices will be complex. These are usually embedded hardware/software devices with many components. Some will include full operating systems and communications capability. These devices are inherently complex. However, the key to minimizing the interaction complexity for the user when dealing with these devices is to hide as much of this inherent complexity as possible. It is not the inherent complexity that makes devices, services, or systems difficult to use. It is the amount of this inherent complexity that is visible to users and with which they must deal.

When there is a large amount of visible complexity, it is difficult for the user to form a clear conceptual model of how the device works. This inability to form a clear, intuitive conceptual model is largely responsible for the difficulty users will have with a device or system [10], [11]. Eliminating visible complexity does not mean trying to eliminate inherent complexity. Indeed, it is likely that as mainstream wearable devices, services, and systems evolve and become more capable they will become inherently more complex. The challenge is to continue to decrease the visible complexity.

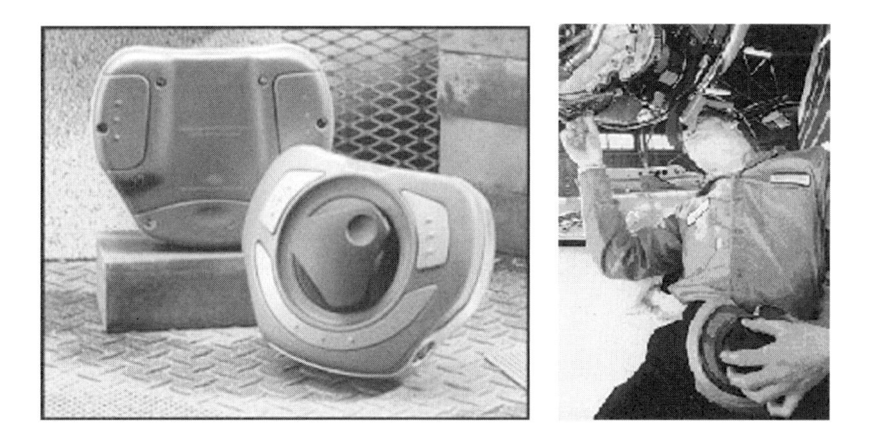

Fig. 5-8. VuMan 3 wearable computer system with novel rotary interface – developed by the Carnegie Mellon University Wearable Computing Group *(CMU Wearables Group)*

Figure 5-8 shows the VuMan3, a wearable computer developed by the CMU Wearable Computing Group [12]. It is quite complex, housing a full computer with wireless capability. However, the external form factor contains a minimum of buttons and a single, large rotary dial that is easy to find without looking at the device. The dial is used to scroll through menu items. The user presses one of the buttons to select the item.

This simple external interface is essential to effectively using the VuMan3 for its intended user population: aircraft mechanics. The user can focus on the aircraft and manipulate the plane's documentation viewed in the heads up display without giving much thought to the use of the computer's external controls.

Yet another way to minimize the visible complexity is to utilize the user's context and individual preferences to offload simple actions and decisions. Another way is to restrict the number of functions in the device or service. One of the primary causes for high visible complexity on today's PCs is that they have so many functions [13].

Fig. 5-9. A Digital Wrist Watch (left) *(Courtesy of CASIO)*

Respect the Inherent I/O Limitations of the Device

Every device has limitations on the type of I/O operations it can effectively support as a result of its inherent design attributes. Failure to respect those limitations or attempting to perform additional operations, will result in a significant increase in interaction complexity.

It is quite possible that the violation of this principle is responsible for much of the increase in Operational Inertia designs incur. While violating this principle can also increase setup effort, we place it in the section of interaction complexity since the user is most likely to experience this effect most often.

As an example, consider a wrist watch. A wrist watch is limited in its support for output by its small screen and small, thin overall size. However, it is even more limited in the type of input it effectively supports due to the lack of space for a keypad or keyboard and limited space for other input affordances such as buttons.

Anyone who has struggled to read the very small day or am/pm annunciators on a feature rich digital watch appreciates the limits of its output capability (see Figure 5-9).

But the real limitation of the watch is with input. There have been numerous attempts at merging PDA functionality with the watch form factor, the latest being the Fossil FX3005 Wrist PDA watch. The PDA functions frequently require alphanumeric input. However, without the use of a keypad or keyboard, such input can be very cumbersome. The Ruputer used a joystick to scroll through a list of characters and select numbers and letters for input. Using a touch screen with Graffiti character recognition, as in the Fossil FX3005, can reduce the effort somewhat, but now you also need a stylus.

None of the watch – PDA hybrids has been widely accepted – from the Ruputer to the Fossil FX3005. The reason is simple: the watch is inherently a display oriented device. Trying to go beyond the limitations inherent in the basic design elements of a wrist watch which make it a display oriented device and attempting to force the ability of alphanumeric text entry into it will inevitably result in a significant increase in interaction complexity[36].

The reason for this is that as soon as you add input capabilities to a watch it goes from being an 'at a glance' one handed device to a high focus, two hand device, a significant increase in interaction complexity.

Many would suggest that the solution is speech input. And indeed, that would be a viable approach if either of the following two conditions holds:

1. the speech recognizer is 100% accurate 100% of the time regardless of the environment and variations in the user's voice brought about by the current activity, health, or other numerous causes of variations in the user's speech; or
2. the mechanism used to correct the speech recognizer when it makes a mistake is swift and easy.

[36] And we should note that this can also increase non-use obtrusiveness. Each of the watch – PDA hybrids has been very large for a watch.

The first condition of accuracy clearly does not exist nor is it likely to in the foreseeable future. The second, the correction mechanism, requires text input in the event that the desired word is not in the suggested correction set, putting us back at square one.

And it not just text input that can cause problems. Watches with cameras, cell phones, and barcode scanners all go beyond the display orientation of the wrist watch, resulting in much higher interaction complexity and often increased non-use obtrusiveness due to their significantly increased size.

Design Command Specifications That Are Consistent and Mutually Reinforcing Across All Input Devices of the Same Modality

A major issue with speech interfaces in a distributed wearable system is command format consistency. If each device implements its own command format, the user must remember the correct way to format a speech command, depending on the device being used. For example, one device may use command and control while another device may use more natural language speech commands. This variation adds significant interaction complexity to the system.

One way of dealing with this is to define a common metagrammar for the system. A metagrammar is a grammar that specifies the structure of commands in other grammars. A portion of the possible metagrammar is shown below. The metagrammar defines the structure of grammars used by any device or application in the wearable system.

Utterance ::= NotificationString (Command | Question | Statement)

NotificationString ::= SystemName

Command ::= CommandPhrase Parameters Information?

CommandPhrase ::= Action Target? Recipient?

Action ::= VerbPhrase

Target ::= NounPhrase

Recipient ::= NounPhrase

Parameters ::= Temporal? Spatial? Condition? State?

These rules specify the structure for the valid utterances that all grammars in the system can define. For example, the first rule says that any utterance must be a command, question, or statement preceded by the name the user gives to the system to get its attention[37]. The subsequent rules shown specify parts of the command structure.

The metagrammar does not specify specific commands, questions, or statements. These are specified by the grammars for each device or application. Nonterminals in the metagrammar that are resolved by the grammars adhering to this metagrammar are called 'deferred nonterminals'.[38] They are shown in regular font in the example above.

Specifying a metagrammar that governs all speech grammars in the wearable system has several advantages:

- user's have to learn only one command structure for all speech enabled devices in the system;
- developers have a higher level of confidence that a speech enabled device or application they add to the system will interoperate with other system devices or applications;
- tools can be developed to automate grammar generation allowing even end users to add speech enabled applications.

All of these help to reduce interaction complexity of devices in isolation as well as for the entire system.

[37] This is necessary if the speech recognizer is doing keyword spotting as an activation mechanism. Such a notification string tells the recognizer to accept the following utterance.

[38] A metagrammar itself may define a few terminals, that is, actual spoken words. These terminals are typically common words used globally throughout the system.

Explanation Based Help
"To create a reminder, say the words, remind me, followed by the time and date of the reminder. Then give the text of the reminder."

Example Based Help
"To create a reminder, follow this example:
remind me at 4 pm on March 22 to go to the store"

Fig. 5-10. Explanation vs. Example Based Help

Utilize Patterns and Analogies

Providing help is important in a mainstream wearable system. Regardless of how simple the mental model and how pervasively it is applied by the designers, there will be times where the user requires help. Speech interfaces in particular pose several challenges for providing effective help. Speech is transient so there is no persistent context to help the user remember where they are in the command process. Speech interfaces are usually used hands free so the divided attention problem is present, and listener fatigue limits the amount of speech to which the user is willing to listen.

The effort required in getting help has a direct impact on the system's interaction complexity. For wearables, which are likely to be used in support of other tasks, the effort required, not just in retrieving help, but making use of it, is crucial. For example, in a speech user interface, once the command for retrieving the help is successfully given, we must make sure that the user can understand the help. Issues of listener fatigue and understandability of the synthesized speech are very important.

One way of dealing with these challenges is to use example based help (Figure 5-10). Example based help gives the user examples of the command rather than describing the parts of the command and how they are put together.

Examples are usually more concise and require less speech to provide. In addition, Cognitive Load Theory (CLT) has shown that examples promote better comprehension of the material with less cognitive load than do detailed explanations [15]. An example provides a familiar context for the information.

Recent CLT research [16] has also shown that as a user's familiarity with the task and expertise in its implementation increases, examples are less of an advantage vis-à-vis detailed explanations. Thus the help system must allow users to obtain detailed help when requested (ex. "Give me details") in the example based help.

Even so, in a wearable system where the user's attention is divided between the wearable and the environment and information must be comprehended rapidly, examples will be an important mechanism for effectively providing help.

To ensure the user gets the help they need, multiple examples should be available so the user can request more than one (ex. "Give me another example"). The selection of examples should cover the most commonly requested help topics[39].

Minimize POLA Violations

Another way to minimize interaction complexity is to adhere as closely as possible to the user's mental model about how the system should function. A key principle for this is to minimize violations of the Principle Of Least Astonishment (POLA). Usually applied to user interface design, POLA says that things should work as the user expects and when elements are ambiguous or conflict with one another the resolution should result in system action that will least surprise the user. The assumption is that the least surprising resolution is the correct one [17].

[39] This may not be the same as the most often used features.

Fig. 5-11. A POLA Violation

In order to minimize POLA violations the system must adopt the user's mental model of the service being provided. This may require supporting extensive customization by the user to allow the service to reflect the user's mental model (at the expense of an initial increase in setup effort).

The system must implement its mental model pervasively. The model should be implemented throughout all aspects of the system. As an example of a POLA violation, we can look at the Windows OS. The Windows OS is built around the desktop metaphor. This metaphor is now understood by all who use a PC. One of the elements of this metaphor in Windows is the shortcut (the Mac OS has a similar element called an alias and Linux has the symbolic link). A shortcut is a document or folder that points to another document or folder. There is no information inherent in the shortcut itself and the shortcut's reference scope is local to the host machine. Clicking (or double clicking as the case may be) on the shortcut takes you to the target. This is a well-understood and frequently used feature.

However, when you click on a shortcut in the Insert Attachment dialog box in Outlook Express under Windows XP Home edition, Outlook Express inserts the shortcut file itself as the attachment (Figure 5-11). In virtually all cases this is not what is desired. The shortcut is useless to its recipient since the reference is invalid across machines. This is an example of a POLA violation. What the user expects is for the target to be accessed and placed in the email message as the attachment.[40]

As mainstream wearable systems and applications become more intelligent they will use software agents that will take actions on the behalf of their user. Many of these actions will be proactive and autonomous, utilizing information about the user, their preferences, and the environmental and situational context [18]. Unless the user has a very good understanding of the mental model embodied in the functionality of the software agent, there can be surprises awaiting the user when the agent does something unexpected.

Another way to minimize POLA violations is to employ standards. Both formal (i.e., developed by recognized standards bodies) and de facto standards can reduce POLA violations. These standards likely embody actions the user is familiar with and will therefore remember more easily.

An interesting example of this is the mobile phone keypad. The 4 x 3 keypad layout and allocation of letters and numbers to the keys is a de facto standard extensively followed by all mobile phone manufacturers almost since the development of the cell phone.[41]

[40] To add insult to injury, when the email is sent, Outlook Express (OE) removes the attached shortcut since it considers it to be an unsafe attachment!

[41] This layout was derived from the Touch Tone phone layout developed by Bell Laboratories in the late 1950s [21].

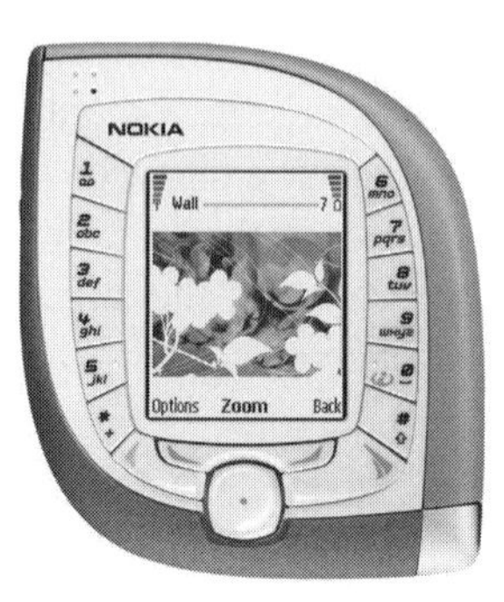

Fig. 5-12. Nokia 7600 phone with nonstandard keypad layout *(photo courtesy Nokia Corp.)*

User's quickly become familiar with the layout and some gain such great expertise with it that they can rapidly enter text using multitap entry without looking. This actually reduces the effort of composing text messages to the point that some users conduct long conversations with one another on their phones.

However, all of this can change when a phone manufacturer modifies the keypad layout. For example, the Xelibri phone from Siemens [19] changes the keypad layout but keeps the keypad below the screen where it is usually found. The Nokia phone (Figure 5 – 12) changes both the keypad layout and its location relative to the display, introducing large variations in spacing between some keys.

Non standard key placement negates use of learned neuromuscular memory. As you repeatedly use the keypad, your mind works with the signals from your fingers to create a memory of the patterns of motion. The more you use the keypad, the stronger this memory of the motions become and the less focus you must apply to the activity to perform it. This neuromuscular memory is the same process at work that allows touch typists to rapidly type without the need to look at the keyboard.

Changing the keypad layout, orientation, or position means the neuromuscular memory associated with the previous keypad geometry is no longer valid. The user must retrain the fingers and mind to create a new neuromuscular memory for the new keypad geometry. This is often more difficult than learning the initial keypad since the previous memory patterns must be 'unlearned' and will sometimes get in the way of learning the new keypad motions. This retraining for a new keypad is a significant source of Learning Operational Inertia.

Disaggregate and Simplify

We discussed the pros and cons of highly physically integrated and modular devices in Chapter 3. We noted that as more and more functions are added to a device, the interactions among the services become more complex. Much of this increased complexity becomes visible in the user interface and the user must deal with it.

In addition, many of these integrated functions have very different input and display requirements, yet they are all forced to use the built in IO affordances of the single integrated device. This results in inevitable compromises in performance and ease of use.

Separating the functionality of the single integrated device into multiple devices, each doing the functions it is best suited for, can result in less visible complexity and better performance and ease of use for each device. Each device has a small set of functions that are highly related to each other, making a clear, simple mental model possible. In addition, each device can provide its set of functions in an optimal way.

For distributed or modular topologies to be viable, a mainstream wearable system must be designed as a system from the ground up. There must be a common architecture for device collaboration. There must also be a common user interface model across all similar devices in the system. This can be very difficult if any device that comes into the range of the wearable's short range wireless network can become part of the system. The wearable system's central unit must provide the unification of the user interface among the devices used in the system. Some of the techniques discussed previously such as specifying a metagrammar to ensure that all speech grammars use the same phrase structure can help to minimize the system interaction complexity that a disaggregated system could generate.

We discuss this issue in more detail in the Chapter 9 discussion on transparent use Multi Modal User Interfaces.

Non-Use Obtrusiveness

One of the distinguishing factors of a mainstream wearable system is the very high level of wearability of each of its elements. This is manifested in their low level of obtrusiveness when they are with the user and not being used for the primary task.

Use Body Conforming Shapes

One of the ways to reduce non-use obtrusiveness is to conform to the body's shapes and motion in all possible planes. Conforming to the shape of the body reduces obtrusiveness and increases user comfort. The conforming mechanism should be flexible so it can adapt to different parts of the body for increased options of device placement. There are several ways to do this. Figure 5-13 shows the evolution of a low obtrusiveness shape.

These shapes are characterized by [20]

- Smooth lines and rounded corners
- Concave inner surfaces to match neutral position body contours
- Convex outside surfaces
- No external, out of form protrusions
- Outside front surfaces slope to rear surfaces at edges

Conforming to the body increases the device's stability on the person since it allows for much more surface area contact with the body. It can also reduce the amount the device extends out from the body, decreasing obtrusiveness. Smooth lines and inwardly curved corners minimize discomfort. It also allows objects that come in contact with the device to glance off of it rather than ensnaring it.

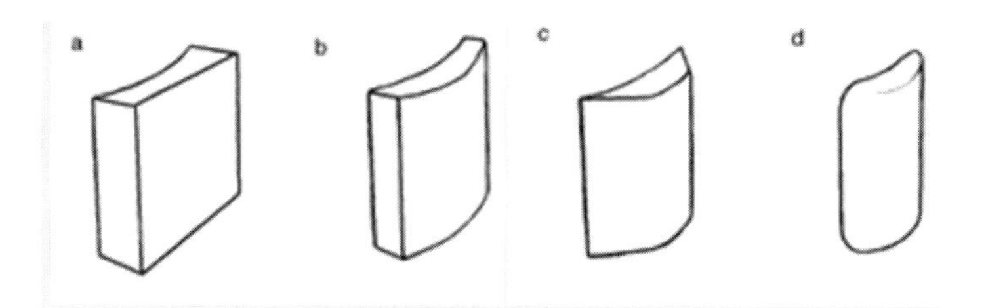

Fig. 5-13. Combining elements of concavity (a) for comfort against the body, convexity (b) on the outside surfaces of the form, tapering (c) as the form extends off the body, and radii (d) softening up the edges combine to create a more humanistic form language and functionality - while creating forms that are physically stronger and designed to deflect (instead of getting caught by) the occasional impacts or pressures against environmental structures/objects (from [20] *(CMU Wearables Group)*

Adapt to Body's Contours and Motion in All Planes

Another method of decreasing non-use obtrusiveness is to increase the accommodation of user movement. This approach has not been seen much, mostly because the devices have gotten smaller, reducing the need for this approach. However, for more complex devices such as the wearable system's central unit and embedded displays this approach can be effective. Accommodating the user's body motion increases the user's inclination to keep the device with them at all times making it available for use with more of the user's tasks.

We can design portions of the device to move independently of each other by including points of flexion on the device resulting in articulated segments. This can be taken further to allow the entire device to flex throughout its form using flexible circuits and overmold materials. Figure 5-14

Fig. 5-14. Segmented Computer Form Factor *(For more information on Infologix Via wearable computer go to www.infologixsys.com)*

shows a Via wearable computer with its distinctive flex joint between the processing module and the storage module containing the hard drive [22].[42]

In the course of a day, the human body can assume a wide variety of positions involving bending, twisting, and reaching. It is important that when we look at using contoured shapes and flexion points we consider all of the positions on the body where the device would be typically worn. Otherwise we may actually limit where the device can be worn because it is shaped such that it is wearable on one area but obtrusive in another.

Do Not Impede Normal Operation of Body Limbs

The device must not impede the motion of the body's limbs when it is worn. We saw that there is an extent of space up to five inches beyond the body that the body takes into account when moving its limbs [11]. We call

[42] Via, Inc was bought by InfoLogix.

this intimate body space. For example, it is what prevents your left hand from hitting your chest when you touch your right shoulder. You are not aware of this space since the body accounts for it automatically. Very thick devices will extend beyond this space and become obtrusive. In extreme cases, the device can become dislodged as the arms knock into it. For example, a thick device worn on the belt can hit the arms as they move back and forth while walking. In many cases, especially when the device can be made flexible, it can be preferable to trade off increased surface area (length and width) for reduced thickness. In addition, body contoured shapes as described above can provide the volume required for the device components while reducing the extension of the device beyond the intimate body space.

In some cases, extension beyond the intimate body space can eliminate or reduce the attractiveness of wearing a device on a specific part of the body. For example, if the device on the wrist is thick, it can impede or prevent the user from reaching into a pocket as the wrist worn device hits the material of the pocket entrance. System designers must take these kinds of issues into account when designing clothing as part of the wearable system.

Reduce Opportunities for Conflict with the Physical Environment

Non-use obtrusiveness extends to the user's environment as well. Conflicts with the environment (ex. getting the device snared in a seat belt, or yanked off the body when hit by something in the environment) greatly increases non-use obtrusiveness. This can cause damage to the device as well as to the garment to which it is attached.

Designers should minimize the opportunities for the device to come into conflict with objects in the physical environment. Using a convex outside surface and rounded corners allows the device to glance off the object it hits rather than become stuck. Protrusions such as extended antennas (see Figure 5-15 left) pose ample opportunities for becoming ensnared with objects in the environment. Mainstream wearable devices should have minimal or (preferably) no protrusions and rounded corners (see Figure 5-15 right).

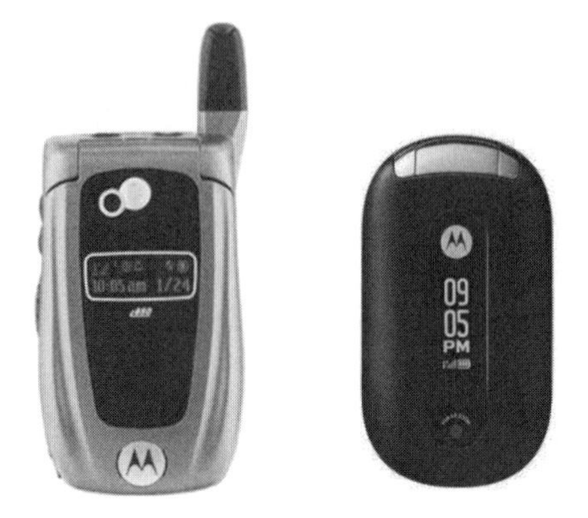

Fig. 5-15. A Potential Candidate for Conflict with the Environment (left) and an unlikely candidate (right) *(Motorola, Inc.)*

Conflict with the environment can also occur when the device is too thick and protrudes too far from the body. This can result in the device banging against walls, door jambs, etc. as the user moves around. This is especially an issue if the user is in a constrained or small space. Trading off increased surface area for decreased thickness is one solution.

Attach To the Body or Clothing in a Non-Invasive, Non-Destructive Manner

Many wearable devices will attach to clothing. If the user is to wear wearable devices all the time, they must be capable of being attached to many different types and weights of garments without damaging the garment or causing discomfort to the user. This attachment must not harm the integrity of the clothing. It should also not distort the shape and line of the clothing. This can limit thc wcight of the device, especially when attached to thin fabric such as some polyester and silk. When attaching to more formal and expensive clothing, the user may not want to use pins or other invasive mechanisms.

Magnetic surfaces under clothing can act as an attachment surface for devices with a magnetic inner surface. The device adheres to the magnetic surface under the garment, requiring no pins or other invasive attachment

$$\text{Output Information Density} = \frac{\text{Information Content}}{\text{Output Size}} \times \text{Relevance to the User's Goal}$$

Fig. 5-16. Components of Output Information Density

mechanisms that might damage the garment. In addition, using a magnetic plate with a large surface area under the garment can securely hold heavier devices to the garment. One issue with this is that the larger plate can distort the line of the garment over it.

When a device incorporates a mechanical button it should be placed on a solid part of the body so that there is a solid base to make pressing reliable. If there is too much travel into the body by the device, the press may not be recorded and the device can become unreliable and uncomfortable. If a solid area can't be used, allow the user to press by pinching the device. This requires the device to incorporate an entrance for the finger or thumb that will stabilize the device and provide the opposition to the press. As an alternative to a mechanical button, a solid-state button can be used requiring no button travel.

To increase the ease of pressing a button without looking at it, a large portion of the front surface area should be the button itself. This is in contrast to many devices such as current Bluetooth headsets where buttons are usually a small portion of the headset's top surface area.

Maximize Output Information Density

When the device is not being used it should not provide the user with information that is not related to the user's task at the moment. The amount of output must be minimized and its relevance to the user's goal must be maximized so as not to distract the user from the primary task. This means maximizing the Output Information Density.

If user notification is absolutely required, the message should specify how the information is related to the user's current task. This will make the intrusion more tolerable and the user will have more incentive to take any action required.

Use of context and techniques such as auto summarization and chunking [1] can reduce the size of the output while maximizing its relevance to the current task. Figure 5-16 shows the components of Output Information Density. The figure is not meant to be taken as a quantitative formula, but rather as an illustration of the relationship of the components.

As an example, many people who carry a cell phone use it to tell the time and no longer wear a watch[43]. A phone without a flip (a 'candybar' phone) usually has a keypad lock function to prevent calls being made inadvertently.

When you want to see what time it is, pressing any button will activate the backlight to see the screen,. However, since the keypad is locked, pressing any key brings up a message telling you how to unlock and lock the phone. Due to the amount of text in the message, it covers up the time (Figure 5-17 left).

Since the phone keypad is already locked, there is only one course of action that can be taken to allow the user to use the keypad: unlock the keypad. There is no point in telling the user how to lock the keypad. Therefore, the Output Information Density could be increased by only telling the user how to unlock the keypad since instructions on locking it are irrelevant to the current task – the support task of activating the backlight in support of the user's primary task of getting the time. This both reduces the amount of output and raises the relevance of the message as a whole to the user's goal, increasing the Output Information Density. This change reduces the size of the message and prevents it from covering up the time, eliminating the entire problem (Figure 5-17 right).

[43] This example and the corresponding pictures are from [23]

Fig. 5-17. Original Message when Keypad is locked (left), Suggested Message (right) *(Photos courtesy of Baddesigns.com)*

5.2 APPLYING THE DESIGN PRINCIPLES

It is often effective to combine these guidelines in designing the mainstream wearable system, resulting in the same or an increased level of wearability. For example, disaggregating the device can result in each piece being less complicated as well as less obtrusive since each piece is smaller.

These mechanisms can be traded off while maintaining or increasing wearability. As stated earlier, the device may be larger if its ability to conform to the body and accommodate user movement is increased.

There is an inherent tension among the components of Operational Inertia. As the size of a device decreases, it will initially become more wearable and less obtrusive. However, beyond a certain point (which varies for each user and device), the device becomes too small and its interaction complexity begins to increase.

Similarly, as a device's size decreases, its in-use obtrusiveness (part of interaction complexity) may also decrease since is becomes easer to handle. Again, making the device too small can actually increase the setup effort because it becomes difficult to manipulate precisely (think about trying to work the small clasp on a piece of jewelry). Only user trials with actual

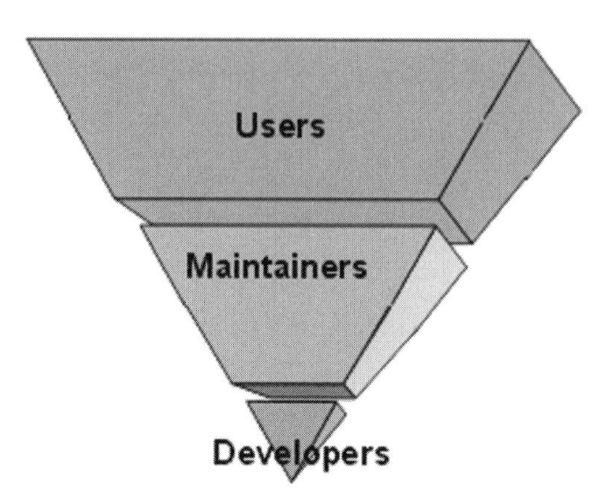

Fig. 5-18. Design Principle Application Hierarchy

devices on a range of users can determine where these design boundaries are.

The Design Principles Application Hierarchy in Figure 5-18 specifies the precedence of the wearable system user, hardware or software maintainer, and hardware or software developer in accepting different solutions embodying the design principles.

User have the highest precedence and solutions should always be chosen that make it easiest for them. Likewise, the system maintainers have precedence over the developers. When it comes to deciding whether to adopt a solution making it easier for the developer or maintainer, the solution that makes it easiest for the maintainer should always be chosen (assuming both solutions result in the same ease of use for the user).

A Design Process

With its emphasis on Activity Based Design and minimizing Operational Inertia, the design process for a mainstream wearable device, service, or system can be a little unusual. The process consists of several steps and it performed in an iterative manner. Note that whenever possible, this process should take place under the actual conditions places, noise, levels, etc. where the devices, system, and services will be used.

1. Define the device's or host's mental model. That is, what are its essential properties, actions, and affordances that make it what it is? This mental model must be intuitive to a user based on general world knowledge and experience. If an object such as a garment is hosting embedded components, the mental model is that of the object without the embedded components. For example, the mental model of a jacket with embedded electronics is that of the jacket without the electronics.
2. Identify augmenting components and their functions. Augmenting components include electronics, processing, communications, etc. that will be embedded in the host device or garment. For each component define its functions and how it interfaces and collaborates with the other components in the host.

   Make sure user actions required by the augmented components either complement or are orthogonal to and not in violation of the user's mental model of the unenhanced device or garment as discussed in Section 5.1.1.

3. List user activities involving the device and their context for all of the user's primary tasks that involve the device. The context of the activity is important since it can define situations or elements that constrain or impede the performance of the activity. Note any such elements or situations. If it is an object hosting embedded electronics, concentrate on performing the functions afforded by the embedded electronics within the normal usage patterns of the host as much as possible.
4. Act out/perform the activities with a model or actual instance of the host object. Carefully recreate the activity within the relevant contexts. Recreate the conditions within the context that impede or constrain the activity.
5. Identify all ZOID violations. That is, identify all incidents of setup effort, interaction complexity, and non-use obtrusiveness in the performance of the activities afforded by the device. Record the impact of each violation.
6. Assign violation severity and priority based on its impact, frequency of occurrence within the activity profile, and availability of workarounds.
7. Cluster violations by OI area (Setup Effort, Interaction Complexity, and Non-use Obtrusiveness).

Design Principle #	Application
Interaction Complexity-2	Lack of controls on surface; Use of contextual controls
Interaction Complexity -7	Separation of display and speaker/microphone from base
Non-Use Obtrusiveness-1, Non-Use Obtrusiveness-2	Concave inner surfaces that face the body
Non-Use Obtrusiveness -3	Trade off additional surface area for reduced thickness
Non-Use Obtrusiveness -4	Convex outer shape, No protruding elements; Flush docking of peripherals
Interaction Complexity -1, Interaction Complexity -5, Setup Effort-3	Automatic activation of peripherals upon undocking; Contextual application initiation upon undocking
Setup Effort -2, Setup Effort -3	Parasitic charging of peripherals while docked

Fig. 5-19. Possible Recording Format for an OI Audit

8. Propose a solution for each violation. List any characteristics of the solution that make it less optimal than the action incurring the violation.
9. Review the effects of the proposed solution on other OI areas. Reducing the OI in one area frequently increases it in another. For example, reducing non-use obtrusiveness by making a device smaller can increase interaction complexity or setup effort due to its smaller size. Repeat the actions adversely affected by the solution and repeat this process with them.
10. Perform an OI design audit. An OI audit identifies each application of the design principles and looks for areas in which additional application of the principles is possible. Then repeat this process with those applications as the focus.

    An example of a possible format for recording the results of an OI audit is shown in Figure 5-19.

The first column identifies the OI principle by its position in the design principle list section (for example, Interaction Complexity -2 refers to 'Overload Normal Maintenances and Activities', the second principle in the Interaction Complexity section of the principles). The second column specifies how the principle is applied to the device under design.

This process is inherently iterative and can require several iterations under the different operating conditions and places before the design is final.

Devices, services, and systems with minimal Operational Inertia will be almost transparent to use. However, there are many other technological issues that must be addressed for true transparency to occur. We discuss the characteristics and challenges of these supporting technologies in the next two chapters.

REFERENCES

[1] Lyons K., Skeels C., Starner T., et. al.., Augmenting Conversations Using Dual--Purpose Speech." User Interface and Software Technology (UIST), Santa Fe, NM, November 2004. pp. 237--246.

[2] Lyons K., Skeels C., and Starner T., Providing Support for Mobile Calendaring Conversations: An Evaluation of Dual--Purpose Speech, Mobile HCI, September 2005

[3] Seiko Kinetic Watches, 2005, http://www.japanese-watches.com/seiko_kinetic.htm

[4] Paradiso, J. A. and Starner, T., 2005, Energy Scavenging for Mobile and Wireless Electronics, Pervasive Computing, IEEE CS and IEEE Computer Society, pp 23-25.

[5] Iso-Ketola Pekka, A Mobile Device as User Interface for Wearable Applications, 2005, http://www.medien.ifi.lmu.de/permid2005/pdf/PekkaIsoKetola_Permid2005.pdf , accessed 2/25/06.

[6] Dvorak J. and Patino J., 2005, Charging system for charging electronic devices in garments, U.S. Patent 6,924,619

[7] Paas F., Renkl A. and Sweller J., (2003), Cognitive Load Theory and Instructional Design:Recent Developments, EDUCATIONAL PSYCHOLOGIST, 38(1), 1–4.

[8] Human Pacman: a sensing-based mobile entertainment system with ubiquitous computing and tangible interaction, Proceedings of the 2nd workshop on Network and system support for games, Redwood City, California, Pages: 106 – 117, 2002

[9] Apple MacBook Design, http://www.apple.com/macbook/design.html, accessed 11/10/06

[10] Norman D. 1989, p12-17.

[11] Norman, D., 1998, pp 154-155.

[12] VuMan, http://www.cs.cmu.edu/~wearable/vuman.html, 1997, accessed 2/20/06.

[13] Norman, D., 1998, pp 78-89.

[14] Berthold, K., and Renkl, A., 2005, Fostering the understanding of multi-representational examples by self-explanation prompts. In B. G. Bara, L. Barsalou, & M. Bucciarelli (Eds.), Proceedings of the 27th Annual Conference of the Cognitive Science Society (pp. 250-255). Mahwah, HJ: Erlbaum, p 251.

[15] Renkl, A., and Atkinson, R., 2003. Structuring the transition from example study to problem solving in cognitive skill acquisition: A cognitive load perspective.Educational Psychologist, 38(1), 15-22.

[16] Cooper G., 1998, Research into cognitive load theory and instructional design at UNSW, School of Education Studies, The University of New South Wales, http://education.arts.unsw.edu.au/CLT_NET_Aug_97.HTML, accessed March 24, 2006.

[17] The Principle Of Least Astonishment, Wikipedia, http://en.wikipedia.org/wiki/Principle_of_least_astonishment, March 8, 2006, accessed March 17, 2006.

[18] Sashima, Akio and Kurumatani, Koichi, Seamless Context-Aware Information Assists Based on Multiagent Cooperation, Cyber Assist Research Center (CARC), National Institute of Advanced Industrial Science and Technology (AIST), http://www.carc.aist.go.jp/~kurumata/papers/AESCS02-sashima-CR.pdf .

[19] Lai, A., 2003, Fashion statements, Computer Times, available at http://it.asia1.com.sg/reviews/phones/por001_20030723.html, accessed March 25, 2006.

[20] Gemperle, et. al., 1998, Design for wearability, Second International Symposium on Wearable Computers, 19-20 October 1998, Pittsburgh, Pennsylvania, USA, pp. 116-122.

[21] Keyboard Trivia, 2006, http://www.vcalc.net/Keyboard.htm

[22] InfoLogix, wearable computers, http://www.infologixsys.com/products/Retail/Products/Wearable-PC/default.asp .

[23] Darnell M. J., 2006, What time is it?, Bad Human Factor Designs, http://www.baddesigns.com/cellphone-lock.html

PART 3: SUPPORTING TECHNOLOGIES

Chapter 6

AWARENESS AND IMMERSION

Making mainstream wearable systems transparent to use requires capabilities not present in current systems. Capabilities such as being intelligent about the user and its environment, providing information in the most effective manner for the user's current situation, and providing sufficient power to all of the system components all assume technologies that are not common in today's wearable systems. This chapter and the next discusses many of these technologies such as pervasive computing, context awareness, and novel power generation and distribution mechanisms and does this through the lens of minimum Operational Inertia.

6.1 PERVASIVE COMPUTING: THE EMERGENCE OF SMART SPACES

In 2004, there were 674 million cell phones [1], 12.5 million PDAs [2], and 177 million PCs [3] sold in the world. However, in that same year there was an estimated 37 *billion* microprocessors, microcontrollers, and DSPs sold[44].

[44] This is based upon 2004 sales of $222 billion [4] and an average price of those chips of $6 [5]. This is a very conservative estimate and the actual number of chips is probably higher.

This comes to about 43 of these chips for every cell phone, PC, and PDA combined. If we count all of the other types of ICs produced, the number is much higher[45].

Where are all of these chips going? Many of them do go into the cell phones, PCs, and PDAs. However, most of them go into other devices such as vehicles, medical equipment, and appliances.

For example, today's automobiles have microprocessors and microcontrollers providing automated control for most parts of the car. ABS and traction control in braking systems, timed interval operation of windshield wipers, multiple zone heating controls, and even self adjusting seats all use one or more microprocessors or microcontrollers [6].

6.1.1 Pervasive Computing Environments

Many of these chips are also going into the home and other buildings to make them aware and intelligent so that they can be an active partner in assisting us with our everyday tasks.

There are many research programs aimed at creating smart spaces. Most of these projects are at universities, although a few companies are pursuing this research as well. Let's very briefly look at some of them and their focus.

Aware Home

The Aware Home Research Initiative (AHRI) [7] at the Georgia Institute of Technology investigates issues of implementing future technologies in the home. It functions as a living laboratory for design, development and evaluation of these technologies. The laboratory is a three-story, 5040-

[45] Other types of chips include memory, RF transceiver chipsets, and low level components such as op amps, inverters, counters, etc.

Fig. 6-1. The Aware Home *(Georgia Tech photo: Gary Meek)*

square-foot home built specifically for the research initiative (see Figure 6-1).

The first and second floors are identical living spaces each consisting of kitchen, dining area, living room, two bedrooms, two bathrooms and an office. The third floor contains several computers used to collect and analyze data, broadband and satellite Internet and campus network access, and other equipment. The basement contains a conference room, storage and machine rooms, and two project work areas.

There are several video cameras and microphones embedded in the first floor ceiling. RFID and other sensing devices are distributed within the rooms including in the floors for presence detection and location tracking.

The research conducted in the Aware Home focuses not only on controlling devices in the house but also on how people interact with a house that has knowledge about where they are and what they might be doing. It also serves as a testbed for several long term research programs including aging in place.

Some of the projects completed or underway at the Aware Home include

- Digital Family Portrait [8]: A picture frame contains a LCD panel and a border with icons. The frame contains an Internet connection that allows it to receive information from a remote location. A typical use is a caregiver using the frame to monitor the activities and status of a remote family member.

  The picture frame receives information about the family member's activity and status and reflects this information in the picture area and its border.

- Cook's Collage [9]: The cook's collage provides a visual summary of recent cooking activity on a kitchen countertop to aid the user in remembering what he was doing in cases of cognitive impairment or interruption. Visual snapshots of the cooking activity are arranged as a series of photos in temporal order, similar to a comic strip. The most recent action is highlighted in yellow. By referring to the snapshots, the user is reminded of the status of the interrupted task, making its resumption easier.
- Memory Mirror [10]: Memory mirror displays activity during a period of time (e.g. 24 hours of a day). Each household item (e.g. medicine bottles, food containers) has a RFID tag on the bottom, and the designated storage area (e.g. medicine cabinet, key tray) has a RFID reader on the top. Each item is photographed and entered into the system's inventory (this is a manual operation). The system tracks the removal from and return to the storage area of each object.

Another example of a smart home is the Innovation Center for the Intelligent House, or 'inHaus", opened in April 2001 in Duisburg, Germany [11]. The house allows researchers to study how people utilize the technology embedded in the house. The house has extensive multimedia cabling and a resident can view any information in any room. Sensors in rooms measure the temperature, air quality and humidity and can automatically open or close windows to keep a stable temperature and conserve energy. Energy is further conserved by the house detecting when all of the occupants have left and then reducing the heating or air conditioning operation.

Fig. 6-2. Elite Care Badge (left), Badge Sensor (center), and Bed Load Sensor (right) *(Components of the Elite CARE System, www.elitecare.com)*

Elite Care

Our last example is not a research program, but an actual, operating, highly instrumented commercial living facility in Oregon. Elite Care [12], [13] is a small family owned business that has created perhaps the most mature pervasive computing environment in existence as of 2006.

The facility is highly networked with sensors that monitor many aspects of the resident's activities and the facility's operation connected to over 30 miles of wiring. Several database servers connected by Gigabit Ethernet compile the sensor readings and analyze trends in each resident's activities and behaviors. Badges worn by the residents send out an infrared beacon detected by sensors deployed throughout the facility to monitor each resident's location and movement.

The sensors (see Figure 6-2) provide data to the monitoring programs that combine several different pieces of low level data to produce a higher level piece of information.[46] This higher level piece of information is more relevant to the staff's duties than the collection of low level data.

We have just discussed a few implementations of a pervasive computing environment. But what are the characteristics of a smart space that ensure pervasive computing truly adds value to a user's everyday life rather than simply screaming technology?

6.1.2 Characteristics of a Pervasive Computing Environment

Mark Weiser, the father of ubiquitous computing[47], stated "The most profound technologies are those that disappear. They weave themselves into the fabric of everyday life until they are indistinguishable from it" [15].

This is more than just making technology invisible. It requires that the technology be transparent to use. The technology within the space must possess many of the same characteristics that a mainstream wearable system does.

The technology embedded into an object such as an appliance, countertop, or door must not violate its primary mental model its users have of its operation. Thus, not only must the technology be invisible but the affordances it provides should be incorporated into those of the host object. This overloads the affordances and activities we normally associate and use with the object to perform or invoke the enhanced functions. This also minimizes POLA violations and does not take away from the experience of using the object.

[46] This process is called data fusion and we will discuss it in detail in the next section on Context Awareness.

[47] Ubiquitous computing and pervasive computing are conceptually very close. .Goff states that pervasive computing would make information available everywhere; ubiquitous computing would *require* information everywhere [14]. However, we will use the two terms interchangeably.

Fig. 6-3. A Mirror with Integrated Display *(All rights reserved Philips Electronics)*

For example, if a refrigerator is instrumented to keep track of its contents and update a grocery list when an item is running out, this function should not require the user to take any other action other than placing the item in the refrigerator and taking it out as needed.

Often new technology enables actions that do not violate the object's typical mental model but add to it in complementary, new ways. In such cases, user intervention dealing with the new functions should be minimized.

An example is shown in Figure 6-3 [16]. Here a bathroom mirror displays several pieces of information. The information is displayed very unobtrusively and remains at the periphery of the user's attention. The user can bring the information to the center of her attention at a glance without interrupting her primary task, in this case, putting on makeup.

Another option of transparent technology is actually being invisible. The technology itself should not detract from the aesthetic qualities of the host object. Many appliances are now designed to be visually compelling in their own right. This often commands a premium in price and any technology that detracts from this will be poorly accepted. Thus the visible complexity of the

Fig. 6-4. Refrigerator with Integrated TV *(LG Electronics USA, Inc.)*

technology must be minimized both to preserve the transparency of operation and to minimize detracting from the object's aesthetics.

Feature integration often results in compromising a mental model. For example, placing a TV monitor in a refrigerator is integrating a function that is not at all what people expect of a refrigerator. And, because it is not expected, the functions do not integrate well with a person's usage pattern of the refrigerator. For example, the TV is often placed on the door opposite, but at the same level as, the ice and drink dispenser, as shown in Figure 6-4 [17].

The level of the ice and drink dispenser is appropriate for its function. However, that same level is inappropriate for the function of the TV. The person standing near the refrigerator must look down at an angle that makes viewing the TV difficult. And the TV is small enough that if you are far enough away from it to see it well, the images are often too small to see clearly for many people.

An alternate approach is to look around the kitchen for an object whose usage patterns and mental model will better accommodate the integration of a TV or computer monitor. One such object is the kitchen counter. We are used to reading newspapers, magazines, cookbooks, etc. lying on the

Fig. 6-5. Display Embedded in a Countertop

counter. Embedding a LCD monitor in the counter with the screen flush with the countertop provides a more natural integration (see Figure 6-5). The normal usage pattern of placing something on the counter to read is overloaded by looking at the monitor[48]. If the monitor needs to be viewed from a distance, it can be raised to a near vertical position and even rotated, much like we prop up a cookbook when we want to read it from a distance.

[48] Of course, the screen would have to be hardened so that it does not violate another aspect of the countertop's mental model – that of being durable. Most countertops are made of hardwood, granite, or laminates such as Formica - all very durable.

6.1.3 Towards Transparent Interaction with Pervasive Computing Environments

There are several challenges for a wearable system interacting with a pervasive computing environment while maintaining the transparent use characteristics described above[49]:

A smart space will typically not become smart all at once. Intelligent devices will be added incrementally. This means the environment will

- not be designed from the ground up to be intelligent. We refer to such a space as an "evolving smart space".

  In an evolving smart space, it is not always clear which devices are smart and which are not. Because of this, the activity of the space can appear unpredictable. That is why it is important that the enhanced activities of a smart object do not violate the user's expectation of how the unenhanced object should act based on the user's world experience with the object - in other words, no POLA violations. This allows the user to use smart objects and dumb objects with minimal confusion and surprise.

  To preserve the transparency of the evolving smart space, the user's wearable could attempt to compensate for the deficiencies in the space when the user or the wearable invokes a capability present in a smarter space but not in the current space.

  If the enhanced function of a smart object would be irreversible, potentially serious in its effects, or incur significant cost or penalty if it

[49] Some of the approaches to minimizing OI given in this section are based on emerging technologies not yet widely available. The goal is to give examples that will illustrate the application of the principles and to start a dialog of ideas on how to apply these principles to achieve transparent use design.

malfunctions, it is imperative that the user be given visibility into the action to be performed and the opportunity to abort the operation.

We can minimize situations like this in several ways:

- Enhanced device functions should be failsoft. That is, they should fail without causing serious effects.
- Each enhanced action by a smart space device should be logged, along with recording the stimuli for the action. The user can review this log upon demand (say on the PC) and adjust device program parameters and preferences to optimize the device's enhanced functions. This is the least intrusive approach, but one that should not be used alone for those actions that have serious consequences, even if they don't malfunction.
- The device should alert the user whenever it is about to take an enhanced action that has potentially serious consequences. The objective is to strike a balance between the goal of transparency and that of control. This is a significant design issue and optimizing it is currently more of an art than a science. In any event, the system should maximize Output Information Density in all of its notifications to the user.

- A smart device can communicate with and respond to the user as well as with other devices in the smart space. In many cases communication with the user will be ad hoc and unplanned since it will be at the whim of the user. If this communication is to be transparent, the wearable system and devices in the space must be capable of impromptu interoperability [18]. This is not just the ability to interconnect, but the ability to do so with little or no advance planning or setup effort. This means that, not just devices, but services, negotiate required capabilities, application data formats, user interface capabilities, and even application semantics.
- As much as possible, the wearable system should push its user interface to the devices with which it interacts. This is another way of minimizing interaction complexity. Because many of the devices within a Pervasive Computing Environment (PCE) will have very simple interfaces, the wearable should push only those elements of its interface that the device is capable of handling. This requires negotiation between the device and wearable system. But whatever interface is provided by the device, it

should utilize the corresponding user interface elements from the wearable system.

- There are a number of reasons why a smart device's function can fail. The software may contain bugs, the device itself can fail in some way having nothing to do with the smart functions, or a required device or service of the environment is not available.

  When something does go wrong with our interaction with a smart space, we must determine where the problem is. So we must become system administrators.[50] We have to read the device logs, either on a PC or on our wearable system and determine what went wrong, where, and how to fix it, or even if it can be fixed.

- Simple maintenance is also an issue. Most dumb appliances don't get upgrades (when was the last time you 'upgraded' an element in your stove?). Other than replacing simple parts such as lamps and cleaning filters, most dumb appliances are maintenance free.

  Smart devices will require software upgrades to fix bugs and enhance security (yes, there will be viruses aimed at smart toasters and ovens). These actions must be transparent to the user to avoid violating the user's mental model of the appliance. This means that, when possible, it must be done over a network.

Table 5-1 summarizes the sources of Operational Inertia for a wearable system in a pervasive computing environment and some possible approaches toward solutions that we discussed above.

[50] This is especially true for early adopters of pervasive computing environments. As smart devices become more widely deployed and their use more common, it is entirely possible that a new occupation will arise – the smart space system administrator. Owners having problems with the performance of their smart space would call such a person to detect and resolve the issues, much like a homeowner will today call an appliance repair person.

Table 6-1. Design Approaches Toward Transparent Pervasive Computing

Operational Inertia Component	Sources of Operational Inertia	Design Approaches Toward Transparent Use
Setup Effort	• Interconnecting devices to the smart space network and other devices	• Auto authorization and authentication based on user preferences, secured information, and passive biometric data stored on the wearable system
	• Configuring a device for opportunistic use	• Wearable system auto-negotiates interfaces, services, and constraints with associating devices
	• Initial user authorization upon network association	• System manages authentication using securely stored user biometric and other data
Interaction Complexity	• User directed information	• Maximize output information density
	• Output of low level data to user	• Combine (fuse) low level data into higher level piece of information relevant to user's primary task
	• Lack of services in current evolving smart space	• Anticipatory caching of information needed in less evolved smart space for seamless transitions
	• Communication among different devices from various manufacturers	• System defines common command and data formats
	• Adjusting to different UIs on opportunistically encountered devices	• Wearable system pushes its interface to the device which uses it for the encounter
Non-use obtrusiveness	• Unsolicited output	• Defer user notification of unsolicited output to user and log it for later review by user

A pervasive computing environment enables our wearable system to interact with our local environment to obtain information about the devices and to control their activities of those devices and services.

There is another, equally important capability for our wearable system: that of obtaining information about ourselves and our situation. This is the area of context awareness and it is to that which we next turn our attention

6.2 CONTEXT AWARENESS

Transparent use requires that our mainstream wearable devices, services and systems be able to make decisions semi-autonomously and even proactively, anticipating our needs and taking action based on our preferences and current situation.

We discussed above how the mainstream wearable system interacts with and controls the environment around us. We now turn our attention on how the wearable becomes aware of and interacts with us. This is the purview of context awareness.

There have been several definitions of context proposed. The one that we will use was proposed by Abowd and Dey [19]:

> "Context is any information that can be used to characterize the situation of an entity. An entity is a person, place, or object that is considered relevant to the interaction between a user and an application, including the user and applications themselves."

This requires that our wearable becomes smart: smart about us, our surroundings, and its own operation and status.

Context awareness involves obtaining the answers to the following questions, listed in increasing difficulty of answering[51]:

1. Where am I?
2. How am I doing?
3. Who/what is around me?
4. What am I doing?
5. Why am I doing the current task?

6.2.1 Where Am I?: Location Awareness

Location awareness is the easiest of the five above questions to answer. There are several technologies, some of them very mature, for determining your location.

The oldest, most pervasive, and most mature is Global Positioning System (GPS). GPS is operated by the Department of Defense (DOD). It uses a constellation of 24 satellites orbiting the earth every 12 hours [20]. The satellites emit continuous precise timing signals that a device can receive and use to calculate its position on the earth.

Basic GPS now provides an accuracy of about 10 meters. The principle limitation of GPS is that the satellites must be line of sight with the device. This means the signals are not well received inside buildings and 'urban canyons'. This greatly limits the use of GPS.

Cell tower triangulation is another location system. Cell towers measure the time a signal takes to arrive from a cell phone. The signals sent out from the phone create a triangle with the three nearest towers. The phone is

[51] Depending on the level of detail and technologies used, the order of questions 2 and 3 in the list could be reversed.

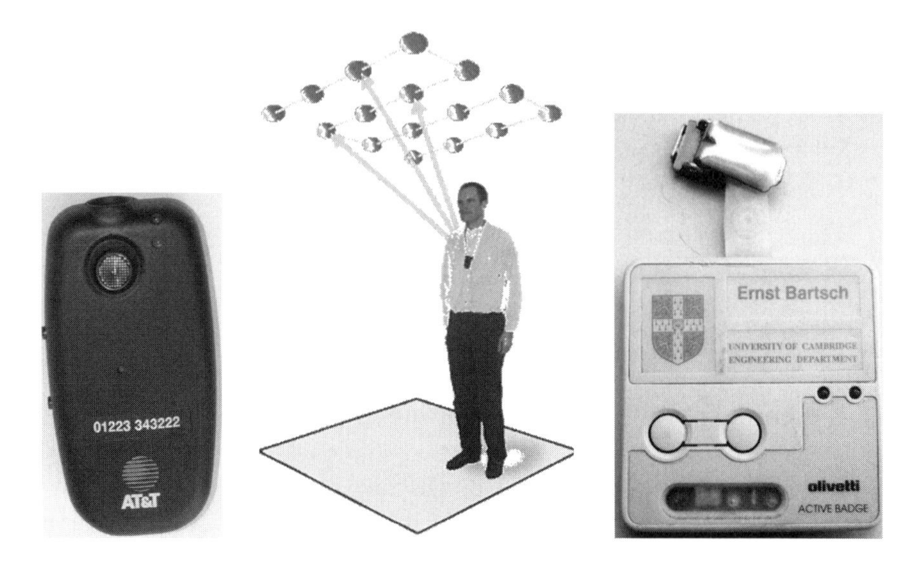

Fig. 6-6. Bat Indoor Location Tracking System *(Photos courtesy of Digital Technology Group, The Computer Laboratory, Cambridge University)*

somewhere in the middle of the three points [21]. Cell tower triangulation can achieve accuracies of only within 100 meters or so.

The inability of GPS signals to penetrate building and the low accuracy of cell tower triangulation has spurred the design of many indoor location systems. The Bat system is an early example. A Bat system consists of a receiver worn by the person and base stations (emitters) placed in a square grid, 1.2m apart on the ceiling of rooms [22], [23].

Each base station periodically transmits a radio message containing a BAT identifier (see Figure 6-6). This causes the addressed BAT to emit a short un-encoded pulse of ultrasound. The base stations in the grid monitor the incoming ultrasound and record the time of arrival of signal from the Bat receiver. The Bat receiver's location is determined based on the principle of trilateration. The system can determine the object's 3D position as well as its orientation. Bat systems are capable of a a spatial resolution of around 3 cm^3.

Fig. 6-7. An AC Powered Location Beacon *(Motorola, Inc.)*

A location tracking system that inverts the tracking information flow is the Cricket system [24]. In the Cricket system objects detect their location by receiving signals from beacons. The beacons are unaware of objects and their location. They simply transmit a RF signal with a location name. Each beacon also transmits an ultrasonic signal. A Cricket receiver measures the time gap between reception of the RF & ultrasonic signals. A time gap of 1 ms roughly corresponds to a distance of 1 foot from the beacon. The receiver chooses the shortest distance as the location. A typical spatial resolution of these systems is less than $1m^2$.

The systems described above are location tracking systems. A defining characteristic of location tracking is that the system can determine the object's location anywhere in the coverage area to within the spatial resolution of the system.

The other type of location system is proximity detection. Unlike location tracking systems, a proximity detection system is not concerned with tracking an object continuously within a space. Instead, it is interested only in detecting when an object is within a specific distance from a discrete set of locations. The absolute location is usually of little or no interest and the distance between the detected locations need not be uniform.

This simplifies the location determination task significantly. It also limits its scope of application. For instance, a proximity detection system may not be comprehensive enough for use in tracking the locations of firefighters in a building. The crucial parameters of a proximity detection system are the density of the beacons and the proximity distance. Of course, if there are a large number of equally spaced proximity beacons the system starts to approach a location tracking system.

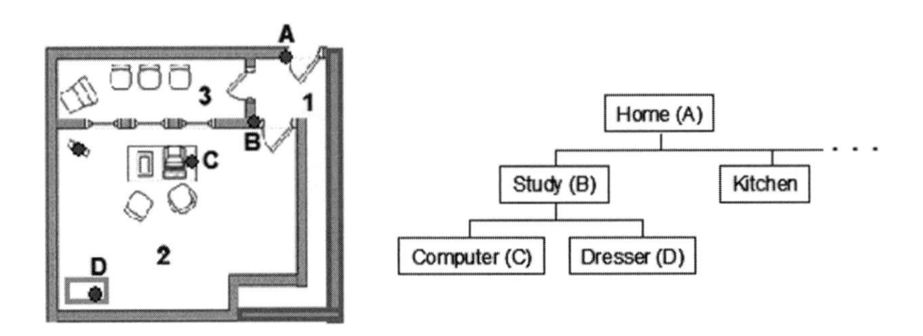

Fig. 6-8. A Space with LoBes (left); Location Hierarchy for the Space right)

An example of a proximity detection system is the LoBe (Location Beacon) system (see Figure 6-7). Each LoBe sends a 128 bit packet with its ID every second at a power level consistent with a 1 meter range in free space to a receiver such as a cell phone. LoBes provide a spatial resolution of 2 meters. LoBes rely on time diversity to avoid collisions. The probability that two or more LoBes within a 2 meter space would transmit in the same or adjacent 25ms windows is quite low.

The wearable device receiving LoBe packets uses the Received Signal Strength Indicator to select the closest LoBe. Each LoBe sends it transmitted signal strength. The receiver selects the LoBe with the lowest transmitted signal strength and highest RSSI to indicate the position. The receiver maintains a list of LoBes and their associated location (set by the user when a LoBe is installed) to resolve the ID of the selected LoBe with the user's current location.

Most indoor location systems have to deal with RF issues such as multipath and interference. Systems operating at frequencies such as 2.4 GHz also have to deal with signal blocking by the human body. All of this means that is it possible that some of the signals will not reach the receiver and cause location updates to be missed.

However, location is hierarchical. If a person enters the room and passes by the LoBe at the room entrance without receiving a signal, the location awareness system can still infer that the user is in the room. For example, in the floor plan on the left in Figure 6-8, LoBe A detects the user is in the house. Suppose that the user enters the Study (Room 2) without detecting the signal from LoBe B. The user's wearable system does not yet realize it is in the Study. However, when the user sits down in front of the computer, the wearable system receives the signal from LoBe C. This signal is associated with the computer. The system's location awareness system maintains the location hierarchy shown on the right of Figure 6-8. The wearable system now knows the user is in the Study since the computer is in the Study.

6.2.2 How Am I Doing?: Biosensing

Maintaining an awareness of the user's condition enables whole classes of applications: personal health, behavior modification, and personal security, to name a few.

There is a wide variety of biosensors, each measuring a specific element of our physiology [25]:

- Electroencephalogram sensors detect brain waves. Experiments indicate that EEG signals can imply emotions such as affection and dislike. However, currently EEG sensors are quite obtrusive and intrusive.
- Respiration sensors measure breathing rate and how deeply the user is breathing. A typical respiration sensor uses a Velcro belt which extends around the user's chest. The band contains an elastic conductive strip that stretches as the subject breaths in and the chest cavity expands. As the elastic strip expands the voltage through the strip changes. This voltage change produces a waveform from which the depth and rate of respiration can be inferred.
- Blood Volume Pulse (BVP) sensors use photoplethysmography to detect the blood pressure in the extremities. Photoplethysmography measures the intensity of light reflected from the skin's surface and the red cells below to determine the blood volume of the respective area.
- Galvanic Skin Response (GSR) sensors measure the skin's conductance between two electrodes. The electrodes apply a tiny voltage across the skin and the resulting current is measured. When a subject is startled or experiences anxiety, there will be a fast increase in the skin's conductance and this increase is measured by the GSR sensor.

- Electromyogram (EMG) sensors measure the electrical activity produced by a muscle when it is being contracted.
- Pressure sensors detect the amount of pressure being applied to whatever the sensor is attached. Solid-state pressure sensors convert pressure over a specified range into a linearly proportional dc voltage or current [26].
- Visual sensors (cameras) capture the image of the user during everyday tasks. Image processing can extract salient features of an image to recognize facial expressions, user movement, and posture.

What can we infer about the user's state given the input from these sensors? This is often a very difficult task, given the variability in each person's response to the same stimulus. Therefore, calibration of each of the sensors for each user is critical.

By the 'state of the user' we can mean many things. At the lowest level, we get direct information about the condition the sensor measures: heart rate, blood pressure, skin conductance, etc. Often readings from two or more sensors are required to provide a definitive identification of the user's state. The combining of multiple sensor data into a piece of higher level information is known as *data fusion* which we discuss in more detail later in this chapter.

Besides identifying physiological states of the user, there is research in *Affective Computing*, giving computers the ability to recognize emotions. This is a difficult task since affective states are internal to the person and involve thoughts as well as physical changes [27]. Some of the physical elements that help express emotion are prosody of speech, posture, gestures, facial expressions, and even pupil diameter [28].

6.2.3 Who/What Is Around Me?

Often the most significant aspect of a user's situational context is the people he is with. Alternatively, in the case of a pervasive computing environment it may be the objects he is near. Such proximity information is often a major influence is what the user does or why he is doing it.

Proximity data, especially about people, who are usually themselves mobile, can be difficult to acquire and/or interpret. Nevertheless, such information can enable innovative applications including location based call routing, facilitating the formation of social networks, [29], [30], [31], and real time event management [32] [33].

Many technologies have been used to determine what and who is near the user. There are two basic mechanisms that can be employed: active and passive. In an active mechanism, the user must take some explicit action to indicate to the wearable what or who is near him. Examples include manually reading RFID tags [34], manually entering the information using speech or text [35], and physically interacting with another person or an object [37].

Passive mechanisms require no explicit action by the user and are thus much less intrusive, although the user has less control over the data input. Examples include communication from an external source such as an active badge [36], [33], using biometrics such as speaker identification [39] or face recognition [38], and automatically reading RFID tags and Bluetooth IDs.

Some of these mechanisms, especially the ones requiring explicit user action, are definitive enough to be used on their own and are suitable for close distance presence detection, such as face to face conversations. For example, using an intrabody communication PAN, each person in the conversation could shake hands as a means of introduction. This would indicate to each user's wearable that it was in the conversation as the wearable received a person's ID when that person shook hands with the wearable's user [37].

Sustained reception of an IR signal would also be an indication that the person sending the signal was standing in front of the receiver and this likely socially engaged with the receiver.

Other presence detection mechanisms are often not sufficiently definitive and must be combined with additional ones to provide results of sufficient accuracy. For example, speaker identification presents several challenges. In order to identify a person by their speech, the user's wearable must have a template of the speaker's voice[52]. This makes identifying newly met people very difficult. In addition, the current state of speaker identification technology requires the utterance spoken by the speaker to be known beforehand by the user's wearable.[53] In addition, requiring people to speak a specific phrase to identify them can be disruptive and unnatural. It adds a level of inertia to spontaneous conversation.

The use of short range RF such as Bluetooth [40] or 802.15.4/ZigBee [41], [42] also has challenges. At 2.4 GHz, signals are significantly attenuated by the body. Thus these systems suffer from some of the same occultation issues as IR, although not as severely. Ranging can be roughly done by using Received Signal Strength Indication, but the accuracy can vary widely with distance and environment.

BlueAware is a system that uses Bluetooth to detect people that are in proximity to the user [31]. The application runs on a cell phone. Each phone receives the Bluetooth Device Address (BD_ADDR) of other cell phones near it. For typical Bluetooth enabled cell phones, this proximity range can be up to 10m. However, as the density of people in the area increases only the Bluetooth signals of the devices closest to the user are not occluded by intervening people and make it to the user's cell phone Bluetooth receiver.

6.2.4 What Am I Doing?

Determining what a user is actually doing can be quite difficult, as many of the indicators (arm movement, posture, vital signs, ambient noise, etc) can be ambiguous and the data is often noisy. In almost all cases the activity must be inferred from these indicators.

[52] In speaker identification the system is not interested in understanding what the speaker is saying. Thus, speech recognition accuracy is not an issue.

[53] However, as mentioned previously, the speaker need not know the utterance beforehand. In fact, to prevent 'spoofing' the system by using a tape recording, it is typical for the person to be identified to be given the identification phrase which they then repeat.

Many different sensors are used in activity recognition research including:

- Accelerometers, 2 or 3 axis, to measure movement and direction
- Temperature sensors to measure ambient and body temperature
- Humidity sensors to measure ambient humidity to gain insight into the weather
- Galvanic Skin Response sensors to gain insight into the person's emotional state. In addition, measuring the onset, peak, and recovery of maximal sweat rates provides insight into evaporative heat loss [43]
- Ambient light sensors (visible and IR) can provide clues of transitions between indoors and outdoors.
- Digital compass to detect orientation and heading
- RFID tags on objects can provide indications of when those objects are handled provided the user wears a tag reader [44]
- Audio such as ambient audio and detection of the user or others speaking
- Barometric pressure can be a good indicator of vertical movement corresponding to an increase (downward movement) or decrease (upward movement) in pressure. The rate of change can also give an indication of the manner of movement. Rapid changes in pressure can correspond to fast movement such as in an elevator while slow changes in pressure can correspond to walking up or down stairs [45].

Most activity recognition systems follow a process consisting of some or all of these steps [46]:

- Acquire the data. This involves getting the data from the sensor to the processing software on the wearable. If the connection between the sensor and the wearable is wireless, a tradeoff might have to be made between the timeliness of the data and the sensor's battery life. The more timely the data must be, the more frequently the sensor must send it across the wireless link, reducing the sensor's battery life.
- Preprocess the data. Sensor data can be very noisy. Preprocessing is often done to reduce the noise. Examples include removing outliers, smoothing the data, and compensating for missing data. One of the first decisions in preprocessing data is to decide on how we will analyze it.

For example, for discrete, low rate data, using a moving average or absolute differences may be appropriate. For high rate or continuous data, analysis in the frequency domain using Fast Fourier Transforms may be more appropriate. The choice of analysis techniques has a significant influence on what kind of preprocessing is done and how.

- Select the most relevant and important features. In systems with multiple sensors there can be a great number of features to analyze. A feature is a property of the data that may be useful for classifying the activity. Examples include the mean, variance, covariance, and pair wise differences between *n* adjacent values. The number of features can be further increased when the data from multiple sensors is correlated.

  In many cases, the effort required to analyze all available features becomes computationally infeasible. In this case it is necessary to reduce the number of features to be analyzed to a manageable set. The objective is to include only those features that have the greatest impact on the variance of the data. This can reduce the original set of features to a manageable size. Some data is lost, but hopefully most of the data variance is explained by the features in the reduced set.

- Extract the features of the data

- Classify the input into one or more activities. The features of the data are used to infer what the user is doing. This can be a single activity such as walking, sitting, or running; or it can be a higher level activity, composed of multiple single activities. Examples are making breakfast, setting the table, or making a phone call [44].

Borriello and Choudhury [45] used several sensors to identify activities including sitting, standing, walking, riding a bike, riding in an elevator, climbing stairs, vacuuming, brushing teeth, and scrubbing dishes. Their data included traces of the microphone input, the x, y, and z axes output of an accelerometer, the magnitude of the accelerometer as a general force vector magnitude, compass heading, IR and visible light, barometric pressure, and UV light.

From these traces we can make out several activities. A high level of audio followed by a lower level corresponds to the subject walking from the outside, where the street is noisy with cars and other loud audio sources, into the building which is much quieter.[54] The barometric pressure trace reveals a gradual decrease with time. This corresponds to walking up stairs. At the same time the compass heading shows a sweep twice through 360 degrees corresponding to turning to go from one flight of stairs to another. A more rapid decrease in the barometric pressure corresponds to riding the elevator up one floor.

The accelerometer magnitude trace shows a large acceleration swing while walking up the stairs. This is characteristic of the acceleration as our legs push us up to the next stair level and rapidly decelerate when we reach the level and our leg straightens out. This pattern happens for each stair. Two smaller and shorter periods of acceleration correspond to the elevator accelerating as it leaves the first floor and decelerating as it reaches the third floor.

All of the data discussed above is fairly low level. The techniques of combining these different pieces of data into a higher level piece of information that more accurately and effectively indicates what the user is doing is called data fusion and is discussed later in this chapter.

[54] However, the audio trace alone would probably not be enough. For instance, we could have moved from a room holding a party (lots of ambient noise) to a quieter room. Analyzing the type of sounds in the audio data could help. For example, if we suddenly no longer hear car horns or the sounds of passing vehicles. Or we could utilize another source such as the sudden loss of GPS (it doesn't work well inside buildings) or receive a beacon whose ID indicates the building's entrance.

6.2.5 Why Am I Doing The Current Task?

Recognizing what a person is doing is only part of the story. Much more insight into their behavior and probable next activities is gained by understanding why they are performing these activities. For example, if my wearable knows that I am setting the table and it also knows it is the time I normally eat dinner, it can infer that I will soon be sitting down at the table and probably will be consuming food. With this information my wearable can schedule other tasks and reminders that are dependent on the fact that I will be eating dinner; for example, I may have medicine that needs to be taken with food. Knowing that I will be eating dinner my wearable can remind me to get the medicine and take it with the meal.

Understanding why a person is doing specific activities requires understanding their goals. Each person has several goals that they establish during the day. A plan is formulated to reach the goal. A goal may be to eat dinner, to visit the doctor, or to go work out. The plan to achieve a goal is composed of tasks. The plan to succeed in working out is composed of getting my workout clothes, going to the gym, and changing into the workout clothes. Each of these tasks is made up of multiple activities. Going to the gym involves grasping my car keys, walking to the car, starting the car, driving the car to the gym, turning off the car, getting out of the car, and walking into the gym.

Recognizing that a person is implementing a plan to achieve a goal is difficult because most plans are not followed straight through [47]:

- People abandon plans before completion. Often this is caused by events or conditions external to and unknown to the context awareness system of the wearable. The user may also forget the plan's goal, i.e. why they are doing the current activities.
- The plan may fail. That is, the plan's tasks are completed but the goal is not achieved. This can also be due to events and conditions that are unknown to the wearable. For example, I may get to the gym but find that the gym is closed for repairs so I cannot work out. This can lead to new goals which may or may not be known to the system.
- People often work on multiple plans at the same time, i.e. they multitask. For example, I may be going through the tasks required to prepare dinner while at the same time doing those tasks required to do laundry. The tasks for the two plans will be interleaved and the wearable's plan

recognition system must be able to recognize this and maintain multiple plan recognition threads.

- Sometimes a task required by one plan will also initiate another plan for a goal the user opportunistically sets because the task was common to both plans. For example, the goal returning home from work involves the task of going into the house. To do this, upon pulling into my driveway, I get out of the car (after turning it off) and start to walk toward the house. However, in doing this I pass the mailbox. The current task of walking to the house subsumes (in this case) the task of walking to the mailbox. Now I adopt the goal to read the mail since it includes the task of retrieving the mail from the mailbox which contains walking to the mailbox. Therefore I walk up to the mailbox, open it, retrieve the mail, close the mailbox and start walking to the house. The plan with the goal to read the mail is now interleaved with the plan with the goal to return home from work.

Each of the cases above makes recognizing plans and their tasks much more difficult.

We have been rather imprecise in our use of the words ‘activity recognition’ and ‘plan recognition’. We now define these terms more rigorously. We will use the following definition from [48]:

> “Plan recognition is a term used to refer to the task of inferring the plan or plans of an intelligent agent from observations of the agent's actions or the effects of those actions. It involves a mapping from a temporal sequence of actions and their effects to an organization of these actions and their effects into some plan representation that identifies the goal of the plan together with the relation between the components of the plan.”

We define activity recognition as analyzing data from one or more sources and combining them as appropriate into a recognition vector that identifies with a high probability the activity the user is currently doing.

Two basic approaches to plan recognition are symbolic, usually entailing some type of formal logic and probabilistic, usually employing some type of Bayesian network.

The symbolic approach has many advantages. The formal theory is independent of the implementation algorithms. It handles concurrent plans, steps shared between plans, and abstract event descriptions.

However, there are some limitations and disadvantages. Perhaps the most serious is that the system can end with no way to select among multiple plans without some specific context information. It may be more effective to model plan inference as a probabilistic process, allowing the system to reason about the relative probabilities of the various models being valid under the current situation, as Charniak and Goldman do in [49].

Charniak and Goldman present a Bayesian model of plan recognition. Their model takes logical axioms as input and builds a Bayes net structure with primitive actions at the leaves and the plan description at the root. After instantiating the values at the leaves of the Bayes net and propagating the values to the root, a token representing the plan is read from the root node.

Bayesian networks are directed acyclic graphs (DAGs). Thc nodcs represent variables of interest such as the ambient temperature, accelerometer data, or the occurrence of an event. The links represent causal influences among the variables of interest. The strength of an influence is represented by the conditional probability that is attached to that parent-child link in the network [50].

A Bayesian network is a model of the causal mechanisms in an environment rather than a model of the reasoning process described in [51]. Bayesian networks allow us to determine answers to queries such as: "Having observed A, what can we expect of B?"[55] and "What is the most plausible explanation for a given set of observations?"[56].

[55] This is known as an associative query and depends only on probabilistic knowledge of the domain.

[56] This is an abductive query and is usually the one asked in plan recognition.

6.2.6 Data Fusion

We have seen that most context awareness systems receive data from multiple sensors and combine this data in some way for classification into a specific activity. This combining data from multiple sensors is called data fusion and can actually be done at different points in the sensing – classification processes[57].

The fusion of data from multiple sensors can be looked at as competitive, complementary, or cooperative [52]. In competitive fusion multiple sensors provide the same data. This redundant information can compensate for degraded reading from one of the sensors and provide more reliable data. In complementary fusion sensors capture different data and do not depend on each other. This provides multiple representations of the activity and provides a broader, more complete picture in which the degraded performance of any one sensor can be compensated for. Finally, in cooperative fusion multiple sensors provide data about the activity in which none of the sensors, taken alone, would be sufficient to identify the activity but the set of data is sufficient to identify the activity with a high probability. Here, however, there is no compensation for the degraded performance of a sensor since the data from each sensor, while necessary, is not sufficient for recognizing the activity.

Data fusion can take place at three points along the activity recognition process. These three levels of data fusion are termed direct, feature level, and decision level. Direct data fusion combines the data directly from the sensors before any feature processing is done. This is most appropriate for arrays of identical sensors where the data being combined has the same features (competitive fusion).

[57] This is also referred to as sensor fusion. We use the broader term of data fusion since we can also combine non sensor data such as time of day, personal calendar data, etc.

Feature level fusion combines feature sets from multiple sensors of different types. This can provide better recognition results since the combined feature set contains more information (features) about the data than either feature set alone.

In decision level data fusion each sensor feature vector is first classified. The results of the classification are then combined. Because the data is combined after recognition, fusion takes place at a semantically high level. This allows the incorporation of domain specific and context based information to be incorporated in the fusion process. At this level, plan recognition can be considered a decision level fusion process.

6.2.7 Toward Transparent Context Awareness

As Figure 6-9 shows, context awareness can involve many stages, each with significant computation and uncertainty. Given such a complex, computationally intensive process that is dealing with inherently noisy and error prone data, the question arises: How can we make the context aware process transparent to the user? To achieve transparent context awareness, we must eliminate the sources of Operational Inertia from the context awareness system and devices.

Setup effort for context awareness of a mainstream wearable system mostly consists of attaching sensors to the body, calibrating them for use, and removing them when taking the wearable system off the body.

Putting the sensors on the body basically boils down to minimizing the effort to attach the sensors to the body and wire them to the wearable system. The most often used approach is to attach the sensors to a tight fitting garment that ensures good contact with the body. This is the approach that the SmartShirt discussed in Chapter 3 takes. Harnesses and single straps have also been used as in the case of ECGs and simple heart sensors. Sensors that do not require close contact with the body such as deformation sensing fabrics and accelerometers are often attached to the outside of clothing such as vests or jackets.

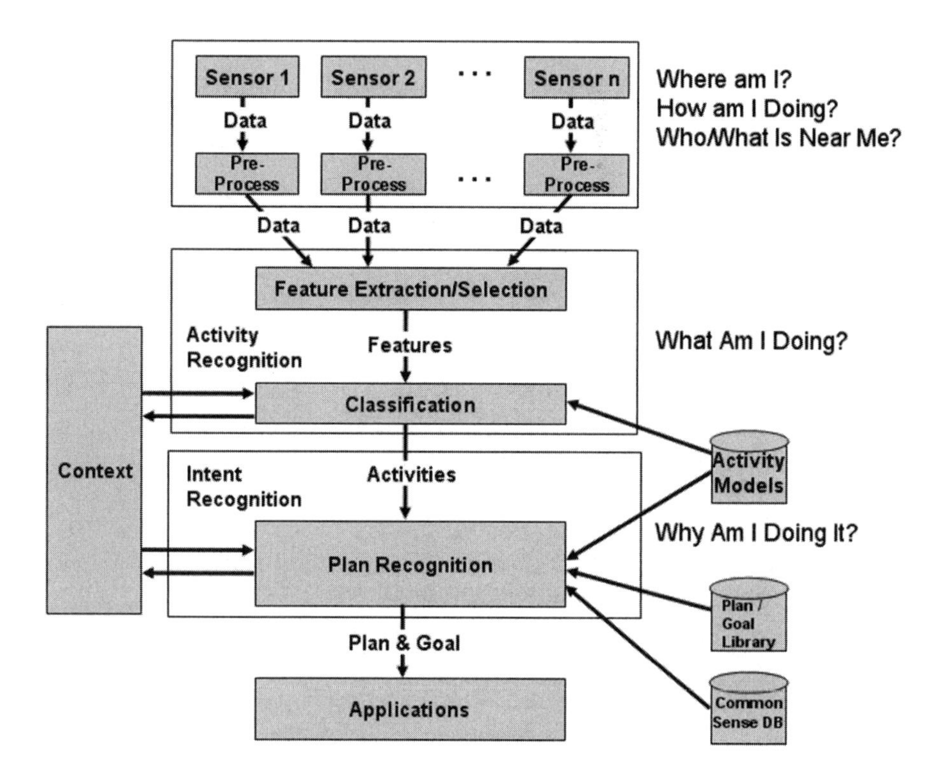

Fig. 6-9. The Context Aware Process

The use of another garment as a vest, shirt, or harness adds setup effort and, depending on the comfort of the garment, can increase in use obtrusiveness (part of interaction complexity) and non-use obtrusiveness. However, it is preferable in most cases to attaching and connecting each sensor individually. Nevertheless, research should continue into other, lower OI approaches.

Many sensors will require calibration. The amount and user intrusiveness will vary with each device. The objective is to use normal body actions as much as possible to calibrate the sensors. This will make the calibration more transparent since the user will not have to go through a separate, explicit calibration activity.

The wearable must take the varying calibration times of the different sensors into account and make sure all of the sensors have been calibrated before accepting data and performing data fusion. Otherwise, it may encounter incomplete data sets as some sensors are still calibrating, making data fusion more difficult and less accurate.

Maintaining these sensors is also part of setup effort. For most sensors this means keeping them powered. Approaches include using normal body actions to trickle charge batteries via parasitic power generation. Examples include arm motion, and heel strikes.

Since the amount of power generated via parasitic generation is small and the sensors may have to operate for an extended period of time, aggressive power management schemes are necessary. If there are multiple sensors required for a specific reading (complementary or cooperative data fusion), schemes such as Groggy Wakeup which seek to power on sensors only when the data is relevant to them can result in savings under certain conditions [53]. This is discussed in more detail in the next chapter.

Finally, if we are using a vest or other garment to host the sensors, the use of power supplying hangers to charge batteries when hanging up the garment (described in Chapter 5) and similar schemes use normal usage patterns of the garment to charge the sensor power sources almost transparently.

For context awareness, the major causes of increased interaction complexity will be caused by data producing incorrect information to the user resulting in the user having to intervene to supply missing or discarded data. Another issue is the system providing data that is not at the proper level of abstraction or is not directly linked to the user's primary task.

One solution is to simply treat late data as missing data for the current update and use past data history of the parameter to estimate the needed value. The use of parameter learning heuristics using expectation maximization (EM) based algorithms to estimate the distribution of the missing data can often compensate for missing data by providing values based on a learned distribution of the data parameter [54].

To help minimize the user's cognitive load, the various pieces of data should be combined via data fusion into a piece of information that is at a level of abstraction consistent with the user's current primary task, is directly relevant to the user's primary task, and is presented in a manner that is concise and understandable and requires minimal user focus and cognitive effort.

Whenever including information that was fused from missing or estimated data it is useful to provide confidence intervals to the applications or user with the information to allow them to decide how best to use the information.

Non-use obtrusiveness of context awareness systems chiefly involves the physical design and placement of sensors. The size and weight of the sensors should be minimized. One way to do this is to separate the sensor from its power supply and use a central power source for all sensors via a power distribution bus. However, the bus itself can also add non-use obtrusiveness if not carefully designed.

Articulated, segmented designs will allow large sensors to conform to the shape and movement of the body and reduce their obtrusiveness. In addition, trading off decreased thickness (depth) for increased height and width can reduce the obtrusiveness of a device and prevent it from coming into contact with the user's moving limbs or elements in the environment.

Placing sensors as close to a pivot point and to the centerline of the body as possible will minimize muscle fatigue since the lever arm supporting them is short. However, this is not always possible since some sensors such as accelerometers sense motion and are best placed far from a pivot point such as at the end of a limb.

Table 6-2 summarizes the sources of Operational Inertia from context awareness in a mainstream wearable system and the approaches toward solutions we discussed above.

Table 6-2. Design Approaches Toward Transparent Context Awareness

Operational Inertia Component	Sources of Operational Inertia	Design Approaches for Transparent Use
Setup Effort	• Putting on and taking off body sensors	• Embed sensors in a tight fitting undergarment so all sensors are applied when

Operational Inertia Component	Sources of Operational Inertia	Design Approaches for Transparent Use
		putting on the garment
	• Calibrating sensors	• Use normal body actions and postures for calibration
	• Maintaining sensor power sources	• Use normal body actions to trickle charge batteries via parasitic power generation • Use power supplying hanger to charge batteries when hanging up garment • Use aggressive power management schemes such as Groggy Wakeup.
Interaction Complexity	• Missing, incorrect, and incomplete data	• Parameter learning heuristics using EM based algorithms to estimate the distribution and value of the missing data [54] • Include confidence intervals reflecting the correctness of data sent to applications
	• Late data due to large sensor latency	• Treat late data as missing data for the current update and use past data history of the parameter to estimate the needed value
	• Data passed to the user that is at the wrong level of abstraction	• Combine relevant pieces of data about the context into a piece of information at the level of abstraction that is consistent with the user's primary task
Non-use obtrusiveness	• Sensor size, weight	• Trade off increased height and width for decreased thickness (depth) • Use articulated, segmented designs for large sensors so they conform to the shape of the body • Separate the sensor from its

Operational Inertia Component	Sources of Operational Inertia	Design Approaches for Transparent Use
		power supply and use a central power source for all sensors via a power distribution bus
	• Sensor placement	• Place sensors as close to a pivot point and to the centerline of the body as possible to minimize muscle fatigue.

Context awareness relies on data transmitted from sensors to the wearable system controller for processing and analysis. We next look at the available and emerging technologies to provide this communication.

REFERENCES

[1] Kim, Ryan, Sales of cell phones totally off the hook 180.6 million units sold worldwide during first quarter, San Francisco Chronicle, May 28, 2005, http://www.sfgate.com/cgi-bin/article.cgi?file=/chronicle/archive/2005/05/28/BUGV4CVUVP1.DTL&type=business

[2] Kort, Todd, et al. Dataquest Alert: Record 14.9 Million PDAs Shipped in 2005, Up 19 Percent Over 2004, Gartner, Inc, http://www.gartner.com/DisplayDocument?doc_cd=137852, February 2006

[3] Krazit, Tom, PC Sales Strong in 2004, PCWorld.com, January 19, 2005, http://www.pcworld.com/news/article/0,aid,119347,00.asp

[4] Therma-Wave., Global Markets Overview, Therma-Wave, Inc, 2005, http://www.thermawave.com/mark_overview.asp#keymarkets

[5] ePanorama.net, Microprocessor and microcontroller pages, March, 2006, http://www.epanorama.net/links/microprocessor.html#general

[6] Yu, Helen, Xilinx FPCs Target Cost-Sensitive Applications, 2/15/03, http://www.xilinx.com/publications/xcellonline/xcell_45/xc_fpc45.htm

[7] Kidd, Cory D., Robert J. Orr, Gregory D. Abowd, et. al., 1999, Newstetter. In the Proceedings of the Second International Workshop on Cooperative Buildings

[8] Mynatt, E D, Rowan, J, Jacobs, A, Craighill, S. Digital Family Portraits: Supporting Peace of Mind for Extended Family Members. Proc. CHI 2001, ACM Press (2001), 333-340

[9] Tran, Q., Calcaterra, G., Mynatt, E. "Cook's Collage: Deja Vu Display for a Home Kitchen." Proceedings of HOIT 2005, 15-32

[10] Memory Mirror, http://www-static.cc.gatech.edu/fce/ecl/projects/dejaVu/mm/ index.html, accessed March 19, 2006

[11] Miller F., Wired and smart: from the fridge to the bathtub, ERCIM News No.47, October 2001, available at http://www.ercim.org/publication/Ercim_News/enw47/ millar.html, accessed March 19, 2006

[12] Stanford V., Using pervasive computing to deliver elder care. IEEE Pervasive Computing, 1(1):10--13, January/March 2002

[13] Ornstein, R., CARE Technology Smart Home System for the Elderly, NIST Pervasive Computing 2001, May 2001,

[14] Goff, Max K., The Scope of Network Distributed Computing, Prentice Hall Professional Technical Reference, May 28, 2004, available at http://www.phptr.com/articles/article.asp?p=170935&rl=1, accessed March 17, 2006

[15] Weiser, M."The Computer for the 21st Century", Scientific American 265(30), pg. 94-104, 1991

[16] Hildebrand, J., Spieglein, Spieglein an der Wand. http://www.gq-magazin.de/gq/5/content/04028/index.php, -August 7, 2003, accessed March 18, 2006

[17] McKay, Gretchen, Smart House: High-tech kitchens let appliances do the cooking, Pittsburgh Post-Gazette, March 19, 2005, http://www.post-gazette.com/pg/05078/473905.stm

[18] Edwards, W. and Grinter, R. (2001). At home with ubiquitous computing: seven challenges. Proceedings of the Conference on Ubiquitous Computing, pp256-272

[19] Dey A. K., and Abowd G. D., (2000), Workshop on The What, Who, Where, When, and How of Context-Awareness, Conference on Human Factors in Computing Systems

[20] Chivers M., (2003), Differential GPS Explained, ESRI, GIS and Mapping Software, http://www.af.mil/factsheets/factsheet.asp?fsID=119&page=2

[21] Smith, A., (2003), Cell-tower triangulation: a workable strategy to improve wireless phones' 9-1-1 access, http://www.fims.uwo.ca/newmedia/newmedia2004/cell/Cell_Smith/cell_smith_triangulation_n28_p.htm

[22] The Bat Ultrasonic Location System, http://www.uk.research.att.com/bat/

[23] Chu, C., 2003, The Anatomy of a Context-Aware Application, CSE 6362 Intelligent Environments Spring 2003

[24] Nissanka B. Priyantha, et al, The Cricket Location-Support System, MIT Lab for Computer Science, http://nms.lcs.mit.edu/

[25] Research on Sensing Human Affect, Affective Computing, MIT Media Lab, http://affect.media.mit.edu/areas.php?id=sensing

[26] Johnson, C. D., 1997, excerpt from *Process Control Instrumentation Technology*, Prentice Hall PTR

[27] Picard, R., 1997, Affective Computing, MIT Press, Cambridge, MA, pp 165-192

[28] Miccoli, L., Bradley M. M., Escric, A., 2004, Looking At The Eye Looking At Pictures, NIMH Center for the Study of Emotion and Attention, University of Florida, http://apsychoserver.psych.arizona.edu/SPRStudent/Awards/2004/lauraposterSPR04.pdf

[29] Terry M., Mynatt E.D., Ryall K., et. al.. 2003, Social net: Using patterns of physical proximity over time to infer shared interests. In Proceedings of Human Factors in Computing Systems

[30] Choudhury, T. K., 2004, Sensing and Modeling Human Networks, Ph.D. Thesis, Massachusetts Institute of Technology,

[31] Eagle N. and Pentland A., 2004, Social Serendipity: Proximity Sensing and Cueing, MIT Media Laboratory Technical Note 580

[32] Cox, D., Kindratenko, V., and Pointer, D., 2003, IntelliBadge^TM: Towards Providing Location-Aware Value-Added Services at Academic Conferences, UbiComp 2003: Ubiquitous Computing, 5th International Conference, Lecture Notes in Computer Science, vol. 2864, pp. 264-280

[33] nTag International – Solutions, http://www.ntag.com/solutions/

[34] The Context Toolkit, 2000, http://www-static.cc.gatech.edu/fce/contexttoolkit/

[35] Rhodes, B. J. 1997, The Wearable Remembrance Agent: a system for augmented memory. Proceedings of the International Symposium on Wearable Computing, IEEE , pp. 123-128

[36] IntelliBadge, Cox D., Kindratenko1 V., and Pointer D., 2003, 1st International Workshop on Ubiquitous Systems for Supporting Social Interaction and Face-to-Face Communication in Public Spaces, 5th Annual Conference on Ubiquitous Computing , Seattle, WA, pp. 41-47

[37] Zimmerman, T. G., 1996, Personal Area Networks: Near-field intrabody communication, IBM Systems Journal, Vol 35, Nos 3&4, 1996

[38] Davis M., Smith M., Canny J. F., et. al., 2005, Towards context-aware face recognition, ACM Multimedia: 483-486

[39] Lin W., Jin R., and Hauptmann A., 2002, Proceedings of AAAI-02 Workshop on Intelligent Situation-Aware Media and Presentation, Edmonton, Alberta, Canada, July 28

[40] Specification of the Bluetooth System, 2004, https://www.bluetooth.org/spec/

[41] Part 15.4: Wireless Medium Access Control (MAC) and Physical Layer (PHY) Specifications for Low-Rate Wireless Personal Area Networks (LR-WPANs), 2003, http://www.ieee802.org/15/pub/TG4.html

[42] ZigBee Specification, 2004, http://www.zigbee.org/en/

[43] Krause A., Siewiorek D. P., Smailagic A., et. al., 2003, Unsupervised, Dynamic Identification of Physiological and Activity Context in Wearable Computing, Seventh IEEE International Symposium on Wearable Computers

[44] Patterson, D.J. Fox, D. Kautz, H. Philipose, M., 2005, Fine-grained activity recognition by aggregating abstract object usage, Proceedings. Ninth IEEE International Symposium on Wearable Computers

[45] Borriello G., Choudhury T., 2005, Activity Recognition: Context Aware Applications, University of Washington Computer Science and Engineering Colloquium Lecture Series, http://www.uwtv.org/programs/displayevent. asp?rid=2658

[46] Nurmi P., Flor′een P., Przybilski M., et. Al., 2005, A Framework for Distributed Activity Recognition in Ubiquitous Systems, IC-AI 2005, pp. 650 – 655

[47] Haigh K. Z., Kiff L. M., 2004, The Independent LifeStyle AssistantTM (I.L.S.A.): AI Lessons Learned, IAAI 04, San Jose CA, July 25-29, 2004

[48] Schmidt, C. F., 2003, Introduction to Plan Recognition, http://www.rci.rutgers.edu/~cfs/472_html/Planning/PlanRecog.html

[49] Charniak, E., & Goldman, R. 1991. A probabilistic model of plan recognition. In Proceedings of the National Conference on Artificial Intelligence

[50] Pearl J., 1997, Graphical Models for Probabilistic and Causal Reasoning, Handbook of Defeasible Reasoning and Uncertainty Management Systems, Volume 1

[51] Kautz, H. A., "A Circumscriptive Theory of Plan Recognition", 1990, Intentions in Communication, Cohen, P. R., Morgan, J. and Pollack, M. E. Eds., MIT Press, Cambridge, MA.

[52] Multi-Sensor Fusion, 2006, Yang G. ed, Springer-Verlag, London, pp 239 – 285

[53] Benbasat, A.Y. and Paradiso, J.A. 2004, Design of a Real-Time Adaptive Power Optimal Sensor System,., in the Proceedings of IEEE Sensors 2004, Vienna, Austria, October 24-27, 2004, pp. 48-51., http://www.media.mit.edu/resenv/pubs/ papers/2004-10-ayb1135v2.pdf

[54] Leray P. and François O., 2005, Bayesian network structural learning and incomplete data. In Timo Honkela, Ville Könönen, Matti Pöllä, and Olli Simula, editors, Proceedings of AKRR'05, International and Interdisciplinary Conference on Adaptive Knowledge Representation and Reasoning, pages 33-40, Espoo, Finland, June 2005

Chapter 7

COMMUNICATION AND POWER

7.1 COMUNICATIONS IN WEARABLE SYSTEMS

Communications is one of the most essential and most resource expensive capabilities of a wearable system. There are several levels of communication that can be present in a wearable system. These include

- Communication through the body for data transfer or to interface with implanted objects [1]
- Body Area Network, typically connecting sensors and other devices on the body with the wearable system's central unit[58]

[58] The term Body Sensor Network is also sometimes used, especially when referring to communication with implanted devices [5]. We choose to use the broader term Body Area Network, to include devices which are not implanted and also non sensor devices (e.g. wearable displays, keypads) as well.

- Person Area Network for communication between the wearable controller and the local environment, and in some cases, directly between wearable system nodes and the environment without going through the system's central unit
- Local Area Network such as 802.11x for communication between high bandwidth nodes in the wearable system as well as between the wearable system and an external LAN access point
- Metropolitan Area Networks such as WiMAX [2] which provide wireless fixed and mobile high speed (up to 40Mbps) connectivity within a 3 km range [3]
- Wide Area Network such as the cellular network for communication with remote parties or devices

We will spend most of our time discussing the Body Area Network.

The term Body Area Network (BAN) usually refers to a network that includes devices worn in or on the body or attached to clothes worn on the body. These devices include implanted sensors, body worn sensors, and non-sensor devices such as small wearable displays, audio headsets, and cell phones. As such, a BAN would be the primary means of communication within the mainstream wearable system.

The characteristics of an ideal BAN and its nodes are many and demanding:

- Because most BAN nodes will be on the body and could be in any possible location on the body, they must be noninvasive and unobtrusive. That means they must be small, conformable, or both. In addition their radiated energy must be at a bare minimum to ensure safety. For example, Specific Absorption Rate (SAR) levels for devices operating within 20 centimeters of the human body, such as cellular phones is 1.6 W/kg[59]. Ideally, the transmit power of BAN nodes would be below the

[59] SAR measures the amount of radio- frequency energy absorbed into human tissue by a radio transmitter.

spurious emission level of electronic equipment like personal computers and portable CD players [6].

- BAN nodes should have a range of about 2m, enough to traverse the height and arm span of most people. A range significantly greater than this consumes more energy and is usually the characteristic of a Person Area Network (PAN).
- BAN nodes should support mesh networking. Each node should be capable of routing messages from one of its neighbors to another. This may be necessary if placement of a node makes communication with the wearable system's central unit difficult due to RF absorption of the body. This is particularly important at frequencies in the 2.4 GHz range and above.
- A BAN should support a significant number of active nodes. Biometric sensors, accelerometers, and other devices will eventually be distributed throughout the body. The number of devices in the BAN at any one time could be significantly greater, for example, than the limit of seven active nodes imposed by Bluetooth.
- Most BAN devices will be very simple in order to achieve their small size and low energy consumption. This also means they must be inexpensive. The BAN transceiver must be a small portion of the total device cost. This means significantly less than US$1.
- For most of the devices in a BAN, energy efficiency is a key element. If there are many nodes, users will not accept having to change batteries, even infrequently. Therefore, alternate energy producing mechanisms such as scavenging from the environment of the body, very short transmission periods with deep sleep modes, and adaptive policies such as Groggy Wakeup [8], are required. Ideally, the transceiver should consume < 3mW during transmission. Supporting mesh networking can also save power since each node need only transmit a very short distance. We discuss the power issue in more detail in the next section of this chapter.
- Many BAN devices only need to transmit data. There is no need to receive. These simplex nodes require much less energy since there is no receiver to power. However, it adds complexity to the BAN as a whole since such nodes cannot support mesh networking. In addition, unless some sort of time slotted or collision detection with back off protocol is

used, the BAN controller must have its receiver turned on constantly to receive the asynchronous transmissions from the simplex nodes.

- BAN devices will be a heterogeneous collection with different data, manufacturers, and purposes. This requires the BAN transceiver and nodes to implement a standard protocol stack.
- Most BAN sensors will transmit very little data and will not require high speed transmission channels. However, there may be some devices such as wearable displays and waveforms such as heart activity that can require up to 2 Mbps or more.

Figure 7-1 shows these requirements for a Ban and compares them to those of other communication mechanisms derived from [6]. The point of the graphs is that the requirements for an ideal BAN are more demanding simply because the need to optimize so many characteristics and the variety of nodes. Note that the obtrusiveness of a BAN and worn sensor networks should be lower than for RFID WPAN/WLAN since the interaction is most likely much more frequent.

So what options do we have for the ideal BAN? For starters, no network technology currently meets all of the requirements of an ideal BAN. This should not be surprising. But how close can we come?

There are several possibilities:

- Technologies for intra-body communication
- Magnetic Induction (not involving body conduction)
- Proprietary radio implementations
- Ultra Wide Band
- IEEE 802.15.4/ZigBee
- Bluetooth

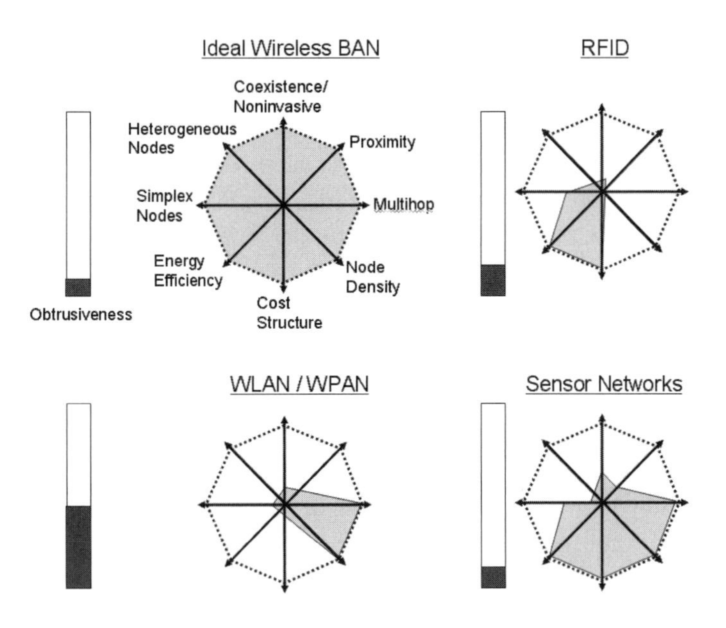

Fig. 7-1. Characteristics of an Ideal BAN compared to other network technologies *(Derived from [6])*

- WiBree

Let's briefly discuss each.

Intra-body Network

Typical networks that use the induction coupling with the body as a transmission medium can only support sensors with small amounts of data sent infrequently. In practice, such networks support transmission speeds of less than 50 kbps.

For example, Skinplex uses the skin as a transmission medium [4]. Small signal generators are worn close to the body to build up an electrical field using a current of 30 nanoamperes which flows across the user's skin while data is being transmitted. Information is transmitted at 195 kHz with a speed of 9600 baud to one or more receivers on the body.

Magnetic Induction

In one way magnetic induction is an ideal BAN technology. RF transmitters generate a modulated RF plane wave in the far field that flows through free space while alternately transferring its energy between its electric and magnetic fields [7]. This radiated field drops of as $1/r^2$ where r is the distance between the transmitter and receiver. In contrast, a magnetic induction system does not rely on the energy radiation for communication. Instead the modulated magnetic field remains relatively localized around the transmitting device. This 'bubble' of communication falls off as $1/r^6$. This rapid fall off in the magnetic field occurs regardless of the presence of metal objects, conductive materials, or people.

The more rapid roll off makes magnetic induction based BANs much less prone to interference and eavesdropping. In addition, the non-radiated nature of the magnetic field means that issues facing RF systems such as nulls, scattering, multipath, fading, and Federal Communications Commission (FCC) limits do not face magnetic induction systems.

The combination of low frequency (11 – 15 MHz), the processing power required, design techniques, and transmit characteristics means lower operating power. For example, the Aura LibertyLink LA116 chipset has an average current consumption of 10 mA supporting a channel speed of 410 kbps across a 1-meter link [7]. By contrast, typical Bluetooth chipsets can consume up to 60 mA (but offer a speed of 1 – 3 Mbps).

Ultra Wide Band

UWB transmitters work by sending very short pulses using a very wide spectrum of frequencies, usually several GHz in bandwidth. The receiver then translates the pulses into data by listening for a familiar pulse sequence sent by the transmitter [9]. As a comparison, narrow band technologies use bandwidth that is typically 10% or less of the center frequency, whereas UWB is defined as any radio technology where the bandwidth is greater than 20% of the center frequency or at least 500 MHz.

The main advantage of UWB over narrowband technologies such as Bluetooth or WiFi is that it provides very high channel capacity using very low power. Shannon's law shows that channel capacity increases linearly with bandwidth while increasing logarithmically in power [10][60]. So by spreading out the signal over a very large area of the spectrum (which in the time domain corresponds to a very narrow pulse), UWB achieves very high channel capacity with little increase in power.

Due to the low power levels mandated by the FCC, UWB devices typically have a range of less than 10m. The FCC requires that UWB radio transmissions operate in the range from 3.1 GHz up to 10.6 GHz, at a maximum transmit power of -41dBm/MHz. Nevertheless, UWB provides very high data rates within these constraints – up to 480 Mbps.

The biggest challenge facing UWB is not technical, but rather political. The IEEE 802.15.3a Working Group, charged in 2002 to establish a standard, dissolved in January, 2006 because the two main industry groups, the UWB Forum and WiMedia Alliance, each representing a different and incompatible UWB implementation, were hopelessly deadlocked. Each group vowed to continue their work on UWB, potentially leading to the two standards battling it out in the marketplace.

IEEE 802.15.4/ZigBee

802.15.4 is an IEEE standard for low cost, low power wireless sensor networks [11]. The standard defines the lowest two layers of a data protocol, the physical (PHY) layer, and the Media Access Control (MAC) part of the Data Link Layer.

ZigBee is an industry driven standard that incorporates 802.15.4 for its MAC and PHY layers and adds networking, interoperability, security,

[60] Shannon's law is $C = BW*\log_2(1+SNR)$, where C = Channel Capacity (bits/sec), BW = Channel Bandwidth (Hz), and SNR = Signal to noise ratio. Further, $SNR = P/BW*N_o$, where P = Received Signal Power and N_o = Noise Power Spectral Density (watts/Hz).

reliable data transfer between source and destination nodes, and other higher level communications capabilities [12]. Although not common, it is possible to use 802.15.4 without ZigBee for very low cost star networks.

ZigBee supports star, mesh (or peer to peer), and cluster network topologies.[61] It also supports beacon and non-beacon enabled networks. In beacon enabled networks, special network nodes called ZigBee Routers transmit periodic beacons to confirm their presence to other network nodes. Nodes typically sleep between beacons, lowering their duty cycle and extending their battery life [13]. In non-beaconing networks, power consumption is very asymmetrical since some devices are always active waiting to receive data, while any others that transmit only can spend most of their time sleeping.

The ZigBee protocol was designed for low power applications. One way it achieves this is to allow non-beaconing networks and allow very long transmission intervals for beaconing networks. Beacon intervals can range from 15.36 milliseconds to over 251 seconds (over 4 minutes) at 250 Kps. However, the current power consumption of ZigBee transceivers during transmission is about the same as Bluetooth transceivers. It is the long duty cycles that give ZigBee devices their very long battery life.

Bluetooth [40]

Bluetooth was originally designed to replace data cables in computers and electronic devices. The protocol specifies requirements that can make it unsuitable for a BAN. For example, the protocol specifies that a Bluetooth network (piconet) can contain at most seven active slaves and one master[62].

[61] A cluster is a group of star networks each terminating on a common overall controller node.

[62] Devices can be in 'park' mode. Devices in park mode do not actively communicate but remain synchronized to the channel and listen for broadcast messages. The master can

Multiple independent and non-synchronized piconets that share at least one common Bluetooth device can be connected in a topology called a scatternet (called clusters in the ZigBee specification)[63]. However the common node can only transmit and receive data from one piconet at a time so communication from the multiple piconets has to be on a time division basis.

The Bluetooth protocol is significantly more complex than the ZigBee protocol. The size of an embedded Bluetooth protocol stack is about 60 Kbytes[42], while a typical full ZigBee stack is < 32 Kbytes and the stack for a Reduced Functionality node can be as little as around 4 Kbytes.

However, the one thing Bluetooth has in its favor is its pervasive deployment. Almost all recently released cell phones have built in Bluetooth. This is a huge advantage for Bluetooth because the cell phone (or whatever it evolves into) is the likely central component of a mainstream wearable system.

Wibree

In October of 2006 Nokia announced a new short range wireless protocol called Wibree[64]. Its main selling points are that it can leverage off of the Bluetooth protocol and it requires much less power than Bluetooth.

---

switch nodes from active to park and vise versa. In this way you can have the illusion of more than seven active slaves. A piconet can have up to 255 nodes in park.

[63] There is a limit of 10 full piconets per scatternet. However, this would translate into 70 nodes in the scatternet, probably enough for most BANs.

[64] The information given here is very tentative. As June 2007, the Wibree Alliance, the trade group promoting Wibree, had not yet officially released the actual spec.

Wibree will use the same PHY (physical) layer as Bluetooth and will thus operate at 2.4 GHz with a power of 0 dbm (Bluetooth Class 2 power level) [14]. It will have a range of 5 - 15 meters (compared to Bluetooth's 1 – 100 meters depending on its power class).

Other aspects of the protocol are simpler than Bluetooth. For example, Wibree devices transmit at 1 Mbps, slower than the Bluetooth EDR rate of 3 Mbps. Also, there is no 'piconet' However, at the time of this writing, it is not clear how many salve devices a Wibree master device can have active. The Bluetooth limitation of seven active slaves in a piconet is one of its main limitations for a BAN.

It requires much less power than Bluetooth. For a standalone implementation (i.e., not collocated on the Bluetooth chip die) in a slave device when transmitting the power consumption is about 15 mA, around 10% to 25% that of a Bluetooth chip. It drops to about 30 μA when in standby mode, and 900 nA in sleep mode [15].

In June 2007 the Wibree forum merged with Bluetooth standards body, ensuring Wibree's close interoperability with Bluetooth. The Wibree specification will become part of the Bluetooth specification and the first version of the specification is anticipated during the first half of 2008.

If the Wibree specification meets all or most of its objectives, it could be a formable option for the BAN of a mainstream wearable system, primarily due to its expected packaging with Bluetooth which is the most widely deployed standard short range wireless protocol available.

Proprietary Radio Implementations

For many sensors in a wearable system, the data acquired will be small (temperature, blood pressure, etc) and does note require high data rates or frequent transmissions (See Table 7-1). In such cases, a proprietary radio implementation, optimized specifically for this kind of sensor data may be most appropriate.

However, proprietary network protocols make it difficult for a BAN to include devices from different manufacturers. Also, since such protocols are not standard, chips implementing them do not enjoy economies of scale, making them more expensive.

Table 7-1. Sensor Data Rate Requirements [16]

Signal	Depth	Rate	Data Rate
Heart Rate	8 bits	10/min	80 bits/min
Blood Pressure	16 bits	1/min	32 bits/min
Temperature	16 bits	1/min	16 bits/min
Blood Oxygen	16 bits	1/min	16 bits/min

How close are we to an ideal BAN? Table 7-2 below compares the protocols discussed.

Table 7-2. Candidates for an Ideal BAN

Characteristic	Intrabody	Magnetic	UWB	ZigBee	Bluetooth	Wibree*
2m Range	N	***Y***	***Y***	***Y***	***Y***	***Y***
Mesh Support	N	N	***Y***	***Y***	N	N
Up To 50 Nodes	?	N	***Y***	***Y***	N	?
< $1US / Node	N	N	N	N	N	?
Power < 3mW	***Y***	***Y***	***Y***	N	N	N
Simplex Nodes	***Y***	***Y***	***Y***	N	N	N
Single Standard	N	N	N	***Y***	***Y***	***Y***
Speed	9.6 kbps	410 kbps	***480 Mbps***	250 kbps	***2 Mbps***	***1 Mbps***

* All figures are tentative as the spec has not yet been released as of June 2007

From the table it is clear that UWB, which has yet to be commercialized, is in the best position to meet all of the needs of the ideal BAN.

Alternatively, we can transfer some of the higher speed, more complex nodes to the Person Area Network, which will most likely be implemented using Bluetooth and use a lower speed, commercially available technology such as ZigBee for the rest. Thus it is likely that we will have at least two short range (< 10m) networks supported by the wearable system.

7.1.1 Towards Transparent Wearable System Communication

To achieve transparent BAN communication, we must eliminate the sources of Operational Inertia from the BAN operation. These include:

- Any authentication of the device with our wearable BAN should be done transparent to the user. Many devices should not require authentication to join our BAN. For those that do, our wearable system must autonomously authenticate the device based on user preferences, secured information, and passive biometric data stored on the wearable system.
- The wearable should manage any issues with limitations of the BAN to minimize intervention by the user. An example is a BAN using Bluetooth which has a limit of seven active nodes. It is very possible that the interaction of the wearable system with the PCE will involve more than seven active nodes. The wearable, to maintain the transparency of the active node limitation of a Bluetooth BAN, must manage the process of placing currently inactive devices into park mode and transitioning other devices from park to active mode transparently to the user.
- At frequencies of 2.54 GHz and higher, the body becomes a very good signal attenuator. This can make it difficult for data from one point on the body to get to the wearable system controller on another area of the body, greatly increasing interaction complexity for the user. One solution is to employ mesh networking and have the node pass its information to a neighbor node which is closer to it and less attenuated by the body. That node passes it to its neighbor that is closer to the system controller

and so on until the data reaches the system controller. While this can reduce the effects of attenuation by the body it complicates the design of the node devices.

- The unavailability of a wide area or local area infrastructure also increases interaction complexity when the BAN must transmit information to or receive information from a remote server. One solution is to have access to multiple communications infrastructures and select whichever one is available and has the strongest signal. Multiband cellular transceivers offer this capability (although the selection is done manually by the user). With the eventual deployment of software defined radio technology, this will be a very attractive approach to ensure the availability of local and wide area infrastructure. When switching from one infrastructure to another, the wearable system must hide this from the applications and users. This may require retransmitting data sent before the switch occurred if the switch happened in the middle of a transmission. This can be avoided by measuring the trend in signal strength of the infrastructure in current use and making the switch before starting a transmission if the signal strength is continually decreasing.
- Receiving information that is not relevant to the current task raises non-use obtrusiveness.[65] Do not notify the user of device or system tasks such as devices joining or leaving the network.

Table 7-3 summarizes the sources of Operational Inertia for wearable system communications and the possible approaches toward solutions discussed above.

[65] Remember, 'non-use' means the user is not using it for their primary task. It does not mean that the system is not using the service or device.

Table 7-3. Design Approaches for Transparent Communication

Operational Inertia Component	Sources of Operational Inertia	Design Approaches for Transparent Use
Setup Effort	• Authentication of new nodes	• Authentication of specific devices by the system without user intervention required for current and predicted tasks
Interaction Complexity	• Body shadowing as the user changes posture and orientation	• BAN supports mesh networking for multiple routing options
	• Lack of communication due to infrastructure unavailability	• Autonomously and transparently switch to an available WAN network via multi-band or software defined radio capability
		• Monitor communication reception trends and perform anticipatory acquisition and caching of required information while communication network is still reliable
Non-use obtrusiveness	• Receiving information not relevant to the current task	• Do not notify the user of device or system tasks such as devices joining or leaving the network
	• Limitation of number of network nodes (ex. Bluetooth)	• System managed active/parked status of network nodes

Communications is one of the most power demanding activities of a wearable system. We now turn to the challenge of providing this power.

7.2 POWER MANAGEMENT

Power is one of the most vexing problems of wearable system design. There are issues of power generation, power storage, and power distribution that currently have very few attractive solutions that enable transparent use of wearable systems.

To better understand why we find ourselves in this predicament, look at Figure 7-2. It shows the improvement in five technologies important to wearable systems relative to 1990 [17]. Disk capacity has been advancing at a pace greater than Moore's law.[66] CPU speed is advancing at the rate of Moore's Law. RAM chip density is increasing at a rate slightly under Moore's Law. Wireless transfer speed, the speed of user information over a wide area network channel, is increasing much slower that Moore's Law.[67] However, if we look at battery energy density, there has been almost no improvement since 1990. It has barely tripled since 1990, while at the same time CPU speeds have increased by a factor of over 1,000.

Although this graph refers to laptop technologies, these graphs accurately represent the state of improvement for wearable systems as well. And the situation is even worse. The rapid increases in disk, CPU, and RAM density have enabled new services such as video, color screens, and fast RISC processors, all of which demand more and more energy.

[66] Moore's law is the empirical observation that the complexity of integrated circuits, with respect to minimum component cost, doubles every 24 months [23]. It is attributed to Gordon E. Moore, co-founder of Intel. Moore's Law is commonly used to measure the increase of processing power over time.

[67] This speed is the speed the user experiences, not the raw channel speed (which can be much higher). The curve for available wireless bandwidth does not include 802.11 hot spots. However, the current density of hotspots is still rather spotty. It should be noted that the curve could sharply turn upward if the emerging WiMax service is pervasively deployed and channel loading remains low. WiMAX is scheduled for nationwide deployment in 2010.

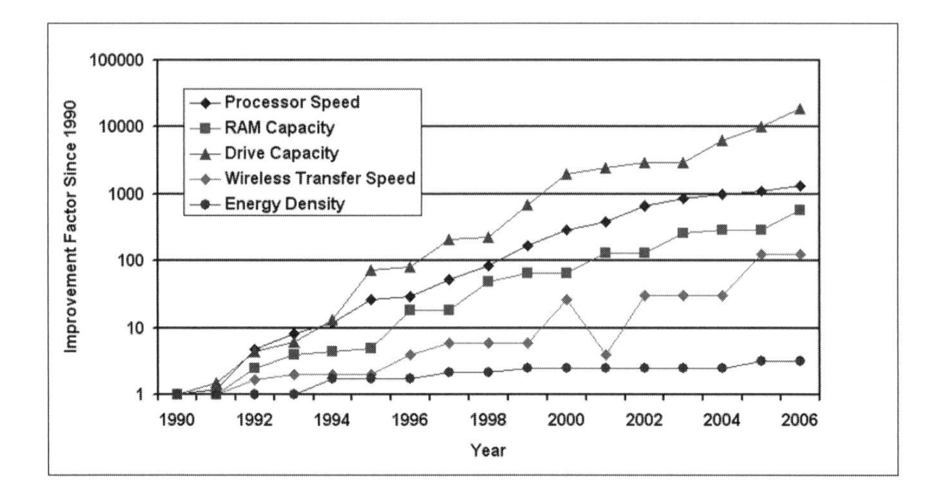

Fig. 7-2. Relative Improvements in Computing Technology From 1990–2006[68]

With the exception of Lithium Ion and Lithium Polymer batteries, there is little in battery technology that we can look to in the medium term to resolve this issue. Thus alternatives must be found.

[68] Unless otherwise noted, figures for 1990 – 2004 are derived from the data in [17]. Single disk drive capacities for 1995 – 2004 from [18]. 2005 single 3.5 in. disk drive capacity from [19]. 2006 single 3.5 in. disk drive capacity from [20]]. Intel® Core™ 2 Extreme Processor performances based on improvements in Composite Theoretical Performance (CTP) scores over CTP score for Pentium III from [21]. 2006 processor performance estimated value based on Intel press releases for 2005 for Dual Core Extreme 8400. 2005, 2006 wireless data rates based on EVDO Rev A. Rates are typical speed per user given nominal loading of the 2.45 Mbps channel shared by all users in a cell [22]. 2005 RAM capacity based on 4 2GB DIMMs on a Dell Precision Workstation 690 from www.dell.com.

7.2.1 Power Generation and Conservation

There are three basic approaches to improving energy availability:

1. Decrease energy usage of the device
2. Use higher energy density sources of stored energy
3. Use continuously available sources of power, such as scavenging energy from the environment and/or the user

Decreasing Energy Usage

Most laptop computer and cell phone designers spend considerable effort to decrease the energy usage of their devices. For example, Intel employs several techniques in hardware and software to lower the power consumption of their Centrino chipsets. These include [24]:

- Lowering the operating voltage of the Processor System Bus
- Dynamically switching the operating frequency of the graphics engine between 200, 133, and 100 MHz based upon the graphics demand of the current application
- Intelligently disabling Dynamic Buffer for both Processor System Bus and memory
- Spinning down a hard drive when there has not been data access for a while
- Using SpeedStep® Technology to optimize power management with multiple frequency/voltage operating points, allowing the processor to drop to a lower frequency and voltage when powered by a battery

Techniques cell phone system designers use in devising optimal power management mechanisms include:

- Dimming or turning off the backlights for displays and keypad
- Shutting off the large color display after a specified interval
- Reducing transmission power to the minimum required by the current distance of the phone to the nearest cell tower

- Powering down unused subsections of the phone. For example, the phone can't ring while a call is in process, so the vibrator circuits can be shut down. Similarly, when the phone is in standby mode the audio circuitry is not needed and can, therefore, be shut down [25].

Higher Energy Density Sources

Despite these techniques, power needs continue to seriously challenge the ability of batteries to keep up. Batteries are a large portion of the weight of a cell phone and often are the leading constraint on reducing the size of the phone. They are also often a large percentage of the weight of the wearable system. Therefore, we must look for sources with higher energy densities.

The leading contenders to replace batteries are micro fuel cells. A fuel cell works by forcing hydrogen gas (H_2) into the fuel cell and through a catalyst on the anode side. When a H_2 molecule comes into contact with the reactant on the catalyst (typically platinum), it splits into two H^+ ions and two electrons (e^-). The electrons are conducted through the anode, and through an external circuit as current and reach the cathode side of the fuel cell. This current is the energy produced by the fuel cell.

Meanwhile oxygen gas (O_2) is forced through the catalyst on the cathode side. The negative charge of the arriving electrons combines with the two H^+ ions arriving at the cathode. Two of these hydrogen atoms combine with an oxygen atom to form a water molecule (H_2O) which is collected as waste.

Fuel cells have 5-10 times the power per unit weight of a Li-ion battery [26], making them very attractive for wearable systems. However, significant technical challenges remain. Most fuel cells are expensive and too large for small cell phones and wearable systems. In addition, the fuel used (typically methanol) is flammable and has not yet been approved by the FAA for use on airplanes. Nevertheless, micro fuel cells currently remain the most viable replacement for batteries.

There are other, more exotic, alternatives. Nuclear batteries are one. Nuclear batteries use charged particle emissions from a radioactive isotope to generate electricity.[69] There are several mechanisms used to convert the radioactive emissions into electricity. One of the most promising is betavoltaics which converts beta particles directly into electricity, much like solar cells convert photons directly into energy. The latest betavotaic cells, called Direct Energy Conversion (DEC) cells, are expected to generate as much as 125 $\mu W/cm^3$, not enough for cell phones or wearable system's central unit but enough for pacemakers and other low power devices [27]. With an expected useful life of over a decade, such batteries would be very attractive for a wearable system's sensors and implantable devices.[70]

However, it is not clear if such batteries will ever be available for consumer electronic devices such as wearable non-medical sensors due to the regulatory and safety certification issues and negative customer perception that nuclear batteries would pose.

Most sensors will require very low power. In fact, the sensing and processing of a node can be a small fraction of the energy expended compared to the wireless transmission of the data, especially if standardized protocols such as ZigBee or Bluetooth are used[71] [16].

The low energy requirements of many sensors have sparked interest in energy scavenging. Energy scavenging acquires energy from the environment or the user.

[69] Nuclear batteries are also referred to as atomic batteries.

[70] The isotope used in the DEC is tritium, an isotope of hydrogen. Tritium has a half life of 12.3 years and 10 % remains after 40 years.

[71] For example, it can take between 100 and 10,000 times more power to transmit a bit of information, even within a BAN, than to execute a single processor instruction [28]

Fig. 7-3. Solar Powered Scott eVest (left) and Accompanying Solar Battery Pack (right) *(photos courtesy of Scott eVest, Inc, www.scottevest.com)*

Scavenging Energy from the Environment

There are several sources in the environment from which to scavenge energy. With all of the RF receiving and emitting devices around us today, the opportunity arises to try to capture and use this ambient RF energy.

However, we immediately see that there are some significant problems. We can approximate the power density a receiving antenna produces as E^2/Z_0, where Z_0 is the radiation resistance of free space (377 ohms) and E is strength of the local electric field in volts/meter [17]. Thus an electric field of 1 V/m yields only 0.26 $\mu W/cm^2$. When we consider the fact that field strengths of even a few volts per meter are rare (except when close to a powerful transmitter), we can see that the power produced by this method is very low.

Even if we deliberately direct RF energy toward a device to power it, as is done with passive RFID tags, we derive very little power. Most passive RFID tags consume between 1 and 100μW. Thus, except for the exceedingly low power sensors, ambient RF energy scavenging is not viable.

Another approach is to derive power from the ambient light. This is exactly what solar cells do. This technique has been applied to wearables – specifically a jacket from Scott eVest (see Figure 7-3) [29]. The removable solar cell panel charges the battery pack which supplies power to the devices in one of the jacket's pockets.

The limitations on solar cells are that it requires a strong source of illumination – preferable direct sunlight. The other limitation is that solar cells are relatively inefficient and produce 100 mW/cm^2 in bright sunlight but only 100 $\mu W/cm^2$ in a typically illuminated office.

There are many other sources of energy within the environment. However, as shown in Table 7-4, none of them produce significant levels of energy.

Table 7-4. Energy-harvesting opportunities and demonstrated capabilities (from [28])

Energy Source	Performance	Notes
Ambient Radio Frequency	$< 1\mu W/cm^2$	Higher near the transmitter [30]
Ambient Light	100 mW/cm2 (direct sunlight) 100 $\mu W/cm^2$ indoor lighting	Typical harvesting efficiencies are around 16 – 20%
Thermoelectric	$60\mu W/cm^2$	At $\Delta T = 5^{\circ}C$. Typical efficiencies are < 1% for T < $40^{\circ}C$ [31]
Vibration/Motion	$4\mu W/cm^3$ Human motion in Hz range 800 $\mu W/cm^3$ for machines in kHz range	For 1 cm^3 generators. [32] Yield is highly dependent on excitation motion
Ambient Airflow	1 mW/cm^2	Demonstrated in a MEMS turbine at 30 liters/min air flow [33]
Push Buttons	50 µJoules/Newton	For the MIT Media Lab pushbutton controller [34]

Energy Source	Performance	Notes
Hand Generators	30 W/kg	Quoted for the shake driven flashlight [28]
Heel Strike	Up to 7 W	Assuming a 1 cm deflection of the piezoelectric material from a 70 kg weight moving at a 1 kHz walk. Actual realizable power of shoe embedded devices is < 1W [35].

Thermoelectric sources create energy from a temperature gradient across an interface. However, usable levels of energy for a wearable system require a significant difference in temperature – over 40°C. This is far greater than those found on the human body or within its immediate environment.

Vibration microgenerators can be placed on walls and floors to capture the low level vibration caused by nearby machinery or human activity. These generators can also be placed on the chassis of vehicles where they can utilize the vehicles vibrations as it travels. However, with the possible exception of vehicles, most sources of vibration are not strong enough to produce significant amounts of energy.

Scavenging Energy from the User

Another source of movement based energy generation is the motion of the user's body, particularly the arms and legs. Self winding watches have been around for some time. These watches utilize a rotary mass mounted off center which rotates when the user's arm moves, generating a current that charges a battery. This energy generation is very small as shown in Table 7-4. However, the energy generation can be increased with more violent motions, as is done with hand shaken flashlights [28].

Another use of hand motion is hand cranked generators and yoyo pull activated generators. These have shown up in radios and flashlights. A yoyo based generator is planned for use in the Hundred Dollar Laptop to allow it to operate in rural areas without electricity [36]. While such mechanisms generate considerable energy, they are hardly transparent mechanisms for creating power as the user must interrupt their primary task to generate energy with these methods.

Use of piezoelectric materials presents another option. When a piezoelectric material is strained, either by being deformed or pressed, an electrostatic potential is created between opposing faces of the material [37].

A person can exert a force of up to 130% of their body weight during the heel strike and toe pressing portions of their walk [38]. Placing piezoelectric materials in the heel and sole of a shoe can generate significant energy as a person walks (see Table 7-4). However, mechanical conversion efficiencies and the constraints of integrating the piezoelectric material and circuitry into a shoe mean that the actual power produced is typically under 1W. Nevertheless, this mechanism generates the most power compared to the others that have the potential to be transparent or nearly transparent to the user.

Pushing a button can also generate power. Pushing a button plunger into a piezoelectric material will strike and deform the material, generating energy. In the Compact, Wireless, Self-Powered Pushbutton Controller developed at the MIT Media Lab [34], enough power was generated with a single button push to power a transmitter for 20 ms at 418 MHz to send a 12 bit code up to 50 feet. However, this method is limited to a short burst of energy without repeated button presses.

There are many other methods of using the human body to generate power including capturing the radiant body heat, the force of breathing, even blood pressure. However, these methods either generate too little energy, are difficult or dangerous to harness, or are very intrusive to the user[72].

[72] For an in depth discussion of these and the other methods of energy generation mentioned in this chapter, see [28]

7.2.2 Transporting the Energy

Besides the generation of energy in wearable systems the other significant challenge is getting the generated energy to where it is needed. For example, although significant energy can be generated using piezoelectric materials in shoes while walking; in most cases the energy is needed elsewhere, perhaps to power accelerometers on the arms or ECG sensors on the chest. Of course, we can always locate the sensors with the major power source, but this is usually impractical.

There are two ways of transmitting power: wired or wireless. Wireless transmission of power is routinely done today in passive RFID tags. A reader sends a pulse of energy to the tag. This energy is inductively coupled to the tag and powers the transmission of the tag's data to the reader. However, without large and powerful readers, such tags can only transmit small amounts of data over short (< 2 in) distances.

Sending larger amounts of power for use by devices with active elements requires higher frequencies, typically microwaves. The power received by a receiver can be expressed as [38]

$$P_r = (G_r G_t \lambda^2 P_t) \ / \ (4\pi R)^2$$

Where:

P_r = received power

G_r = Receiver gain

G_t = Transmitter gain

λ = Operating wavelength of the antennae

P_t = Transmitted power

R = Distance between the receiver and transmitter

Power increases as the square of the wavelength but decreases as the square of the distance[73]. Another advantage of higher frequencies such as microwaves is that they can be focused much more narrowly, increasing the amount of power actually received by the receiver antenna.

It is interesting that, even though the received power decreases as $1/R^2$, power decreases in transmission lines as $e^{-2\alpha z}$, where α is the attenuation constant of the line and z is the length of the line [38]. This means that, for long distances, wireless power transmission can, theoretically, be more effective than wired transmission.

However, besides the free space propagation loss, represented by the $\lambda^2/(4\pi R)^2$ term in the equation, we must account for attenuation caused by the body and other objects that can come between the receiver and transmitter in a wearable system. When we account for this object attenuation and also the multipath effects the received power decreases as $1/R^n$, $n > 2$ [39]. The actual value of n depends on the type of objects causing the attenuation effects and the signal frequency[74].

When dealing with wireless transmission of power we must address the issues of safety, whether real or perceived. Microwaves, which include the popular 2.4 GHz frequency, penetrate the body and heat it. In high enough

[73] The above equation specifies the maximum power that can be received by the receiver. Many factors including impedance mismatch at either antenna, multipath effects, and additional attenuating effects all work to make the actual received power less than that specified by the equation.

[74] The body, for example attenuates signals at higher frequency better that those of low frequency. At 2.4 GHz, a common frequency for BANs and PANS, the body is a very effective signal attenuator.

levels, this can be dangerous, particularly in the areas of the brain, gentiles, eyes, and stomach [38].

The IEEE has defined Specific Absorption Rate (SAR) limits for radiating sources which are typically near the human body. The limits for the 100 kHz – 6 GHz range, which includes most of the PAN and BAN networks used for wearables, are given in Table 7-5.

For wearables, this can be a significant issue since these devices are close to the body for significant periods of time and can be placed on many different areas of the body. In addition, substantial shielding is usually not an attractive option as it makes the devices bulkier, heavier, and more rigid.

Table 7-5. IEEE SAR limits for 100 kHz - 10 GHz (from [44])

Exposure Characteristics	Frequency Range	Whole Body Average SAR[3] (W/kg)	Localized (Head and Trunk) SAR[4] (W/kg)	Localized (Limbs) SAR[5] (W/kg)
Occupational Exposure[1]	100 kHz – 6 GHz	0.4	8	20
General Public Exposure[2]	100 kHz – 6 GHz	0.08	1.6	4

Notes

[1] Exposure averaged over a 6 minute period

[2] Averaging time varies from 6 minutes to 30 minutes

[3] Averaged over the entire body

[4] Averaged over 1g of tissue in the shape of a cube

[5] SAR for hands, wrists, feet, and ankles is averaged over 10 g of tissue in the shape of a cube

Fig. 7-3. Constraints of Wired Power Distribution

Using wires to transmit the power from its source to where it will be used presents its own set of challenges. Most significant is that wires can constrain the user's movements and are a source of failure.

Referring to Figure 7-3, we see one of the issues. The user wears shoes that generate power as he walks via piezoelectric inserts. However, the power is needed in the user's shirt to power several sensors. The pants contain a wired power distribution system that can be used for any device embedded in the pants. Similarly, the shirt contains a wired power distribution system to deliver power to all devices embedded in it. However, how does the power in the shoe get to the pants and from the pants to the shirt?

One possible mechanism is to use conductive fabric to create a contact bridge between the shirt and pants. For example, the inside surface of pants at the belt area is lined with conductive Velcro fabric. This fabric is hooked to the power distribution network of the pants which is connected to the power generator of the pants.

The outside surface of the shirt contains conductive strips that run vertically from the bottom of the shirt up to some level such that there is a high probability that they will come into contact with the pants inner conductive liner at the belt level. The strips are connected to the power distribution network of the shirt.

When the user wears the shirt tucked into the pants, at least one of the conductive strips of the shirt will be in contact with some area of the conductive inner lining of the pants and connect via the Velcro adhesion mechanism. The pressure induced by the person's body and/or the fastened belt will insure a good contact between the shirt strip and the pants lining. In addition, the length of the vertical shirt strips, their multiplicity, and the fact that the pant's inner conductive liner extends around the entire circumference of the pants provides a high probability of continued contact even as the person moves and assumes many different postures during the course of the day.[75]

Bridging the shoe – pants gap is much more difficult since the pants do not typically come into sustained contact with the shoes.

Wires running through garments are also a source of failure. As discussed in Chapter 5, usage patterns for electronics enhanced garments should not differ from those of unenhanced garments, even to maintain and use the embedded electronics. This means the user should be able to wash, iron, fold, and even crumple up the shirt. This requires the embedded wires to be strong, but flexible. It may also require multiple paths for communication and power distribution to accommodate single path failures due to a broken wire.

[75] A serious issue with this mechanism which may make if non-viable is the potential for harm to the user if the conductive fabric of the garment bringing the power (the pants in the example) comes into contact with the user's skin.

7.2.3 Towards Transparent Power Management

To achieve transparent power management, we must eliminate the sources of Operational Inertia (set up effort, interaction complexity, and non-use obtrusiveness) from power generation, distribution, usage, and maintenance.

For power management in a mainstream wearable system setup effort mainly involves attaching, removing, and maintaining the system's power source (or sources), in most cases batteries.

In some cases such as electronic devices highly embedded in garments or other hosts, the batteries must be considered permanently embedded. In that case they must be protected from operations in the normal life cycle of the host that could damage them, such as washing and ironing in the case of garments. Waterproofing them via enclosing them in silicone overmolds or cases is one possible approach.

If the batteries are not to be permanently embedded, then they must be easily and rapidly removable and replaceable without damaging the host. Minimizing this process reduces non-transparency since these actions are most likely not in keeping with the mental model and expected interaction with the unenhanced version of the host (for instance a jacket). The battery or battery cluster attachment points to the system's power bus should be minimized. Attachment mechanisms should require minimal focus. Examples include insertion guides (magnetic or physical), and press-and-release mechanisms for easy removal.

For embedded electronics we seek to use the normal usage patterns of the host to charge or maintain the power sources. Examples include using kinetic motion, walking and using the striking power of the heel, and using power carrying hangers and rods to charge batteries embedded in garments.

Setting up bridging mechanisms between garments to transfer power from one garment to another is difficult to do transparently. Conductive Velcro strips on garment pairs that typically come into contact with one another is one possible approach, although its effectiveness can vary widely depending on where this mechanism is applied. There are currently few alternatives other than explicitly inserting a jack from the power bus of one garment into the plug of the power bus of the other garment.

Power generation contribution to the system's interaction complexity stems mainly from power depletion or interruption while the wearable system is being worn. To minimize the effect of broken power distribution paths, multiple paths may be necessary.

Aggressive power management using the user's activity profile and context along with incremental activation policies such as Groggy Wakeup can lengthen the operating life of the power source. Using parasitic power generation techniques such as piezoelectric inserts into footwear can provide some energy to extend the life of the power source while maintaining a high degree of transparency.[76]

Providing a common power bus to many of the wearable system's components can reduce the size and weight of each component, and thus its non-use obtrusiveness. In addition, segmented battery clusters that conform to the body's contours can reduce obtrusiveness.

Thc usc of a common power distribution bus requires running insulated wires throughout the host's inner surface, such as a jacket. The wires need to be highly flexible to withstand the treatment commonly associated with its unenhanced version. For example, a shirt is typically thrown in the hamper or folded. These actions can place significant stress on embedded wires. The wires need to be highly bendable and somewhat stretchable since the shirt can experience stress in any direction as the user handles it.

Surrounding heat producing elements with heat conducting and phase changing materials to draw heat to the host's surface will reduce potential user discomfort, and with it, non-use obtrusiveness

Table 7-6 summarizes the sources of OI for power management along with possible approaches for solutions that we discussed above.

[76] However, the piezoelectric inserts may stiffen the shoe, causing the user discomfort while walking and adding to in-use obtrusiveness.

Table 7-6. Design Options for Transparent Power Management

Operational Inertia Component	Sources of Operational Inertia	Design Approaches for Transparent Use
Setup Effort[77]	• Removing and replacing batteries	• Sealing batteries inside silicone overmold for semi-permanent embedding • Single point of connection of batteries to power bus. • Easily accessible battery storage within host for easy removal and replacement
	• Charging batteries	• Overload normal host maintenance activities. For example, use kinetic motion, use power transferring clothes hangers, etc.
	• Bringing power across garments	• Inter garment power bridging techniques to transfer power across pieces of clothing (ex. Pants and shirt) using conductive Velcro or physical jacks and plugs
	• Attachment/removal of devices to/from power distribution	• Quick, eyes free attachment mechanisms (such as press and release, magnetic guides, etc)
Interaction Complexity	• Power distribution failures during use	• Multiple power paths for fault tolerance in case a wire breaks
	• Inadequate battery life	• Context based system activation uses power only when needed and only as much as required (ex. Groggy

[77] Setup effort includes maintenance effort

Operational Inertia Component	Sources of Operational Inertia	Design Approaches for Transparent Use
		Wakeup)
		• Trickle charge batteries with energy scavenging sources
Non-use obtrusiveness	• Impact of power supply on device size and use	• Power bus to distribute power to devices from centralized source
		• Segmented battery clusters that conform to the body's contours
	• Excessive heat generation	• Surround heat producing elements with heat conducting and phase changing materials to draw heat to host's surface
	• Obtrusiveness of wires for power distribution	• Flexible, stretchable wires to accommodate user motion and postures

There is another set of technologies essential to a wearable system. The technologies employed in the system's user interface will in large part determine the quality of the user's experience with the system. We next discuss those user interface technologies.

REFERENCES

[1] Zimmerman T. G., 1996, Personal Area Networks (PAN): Near-Field Intra-Body Communication, IBM Systems Journal http://www.research.ibm.com/journal/sj/353/sectione/zimmerman.html

[2] IEEE Std 802.16e™-2005, 2005, Part 16: Air Interface for Fixed and Mobile Broadband Wireless Access Systems, IEEE, http://standards.ieee.org/getieee802/download/802.16e-2005.pdf

[3] WiMAX Technology, 2005, WiMAX Forum™, http://www.wimaxforum.org/technology

[4] Ident Technology AG, 2005, Skinplex http://www.ident-technology.com/index.php?option=com_content&task=blogcategory&id=3&Itemid=4&lang=en

[5] Higgins H., 2006, Wireless Communications, Body Sensor Networks, Guang-Zhong Yang ed., Springer, Boca Raton, Fl, pp. 117 – 122

[6] Noninvasive Wireless Body Area Networks, 2005, Wireless Communications Group, ETH Zurich, http://www.nari.ee.ethz.ch/wireless/research/projects/ban.html

[7] Aura, Inc., 2005, http://www.auracomm.com/site/content/roll_off.asp

[8] Benbasat, A.Y. and Paradiso, J.A. 2004, Design of a Real-Time Adaptive Power Optimal Sensor System,., in the Proceedings of IEEE Sensors 2004, Vienna, Austria, October 24-27, 2004, pp. 48-51., http://www.media.mit.edu/resenv/pubs/papers/2004-10-ayb1135v2.pdf

[9] Ultra Wideband (UWB) Technology, 2006, Intel Technology and Research, http://www.intel.com/technology/comms/uwb/

[10] Wilson J. M., 2002, Ultra-Wideband / a Disruptive RF Technology?, Intel Research & Development, Version 1.3, www.intel.com/technology/comms/uwb/download/Ultra-Wideband_Technology.pdf

[11] 802.15.4 Specification, 2004, Wireless Medium Access Control (MAC) and Physical Layer (PHY) Specifications for Low-Rate Wireless Personal Area Networks (LR-WPANs)

[12] ZigBee specification, 2004, ZigBee Alliance, http://www.zigbee.org/en/spec_download/download_request.asp

[13] ZigBee , 2006, Wikipedia, http://en.wikipedia.org/wiki/ZigBee

[14] Prophet G., 2007, Wibree – wireless PAN with long battery life, EDN Europe, June 14, 2007, http://www.edn-europe.com/wibreewirelesspanwithlongbatterylife+article+405+Europe.html

[15] Wibree specification takes shape, The Wibree Quarter, Quarter 1 2007 www.wibree.nordicsemi.no

[16] Body Sensor Networks, 2006, Yang, Guang-Zhong (Ed.), Springer, London, pp 184

[17] Paradiso, J. A. and Starner, T., 2005, Energy Scavenging for Mobile and Wireless Electronics, PERVASIVE computing, IEEE Computer Society., Volume 4, Number 1, pp 18 – 27

[18] Historical Notes about the Cost of Hard Drive Storage Space, http://www.littletechshoppe.com/ns1625/winchest.html, accessed May 31, 2007

[19] Digest 2005: Hard Disk Drives and SATA/SAS Controllers, http://www.digit-life.com/articles2/storage/itogi2005hdd.html, accessed May 31, 2007

[20] PC Connection, http://www.pcconnection.com/ProductDetail?Sku=6757549, accessed May 31, 2007

[21] Intel Processors CTP Calculations, http://www.intel.com/support/processors/sb/CS-022818.htm, accessed May 31, 2007

[22] Airvana, CDMA2000 1xEV-DO Overview, http://www.airvananet.com/technology/technology_evdo_rev_a.htm, accessed May 31, 2007

[23] Excerpts from A Conversation with Gordon Moore: Moore's Law. , 2005, Intel Corporation , pp 1

[24] Optimizing Battery Life, 2005, Intel® Centrino® Mobile Technology, http://www.intel.com/support/notebook/centrino/sb/cs-009877.htm

[25] Hurtz G., 2003, CDMA handset design challenge: 11 separate power supplies, CommsDesign, http://www.commsdesign.com/design_corner/showArticle.jhtml?articleID=12804119

[26] Fujitsu Develops High Capacity Micro Fuel Cell Technology, 2004, Fujitsu, http://pr.fujitsu.com/en/news/2004/01/26-1.html

[27] Direct Energy Conversion Technology, 2004, BetaBatt, Inc, http://www.betabatt.com/index.html

[28] Starner T. and Paradiso J.A., "Human-Generated Power for Mobile Electronics," Low-Power Electronics Design, C. Piguet, ed., CRC Press, 2004, chapter 45, pp. 1–35.

[29] Scott eVest Inc, 2005, Solar SeV, http://www.scottevest.com/v3_store/access_solar.shtml

[30] Yeatman E.M., "Advances in Power Sources for Wireless Sensor Nodes," Proc. Int'l Workshop Wearable and Implantable Body Sensor Networks, Imperial College, 2004, pp. 20–21; www.doc.ic.ac.uk/vip/bsn_2004/program/index.html.

[31] Stevens J., "Optimized Thermal Design of Small ΔTThermoelectric Generators," Proc. 34th Intersociety Energy Conversion Eng. Conf., Soc. of Automotive Engineers, 1999, paper 1999-01-2564

[32] Mitcheson P.D. et al., "Architectures for Vibration-Driven Micropower Generators," J. Microelectromechanical Systems, vol. 13, no. 3, 2004, pp. 429–440.

[33] Holmes A.S. et al., "Axial-Flow Microturbine with Electromagnetic Generator: Design, CFD Simulation, and Prototype Demonstration," Proc. 17th IEEE Int'l Micro Electro Mechanical Systems Conf. (MEMS 04), IEEE Press, 2004, pp. 568–571.

[34] Paradiso J. and Feldmeier M., "A Compact, Wireless, Self-Powered Pushbutton Controller," Ubicomp 2001: Ubiquitous Computing, LNCS 2201, Springer-Verlag, 2001, pp. 299–304.

[35] Antaki J.F. et al., "A Gait-Powered Autologous Battery Charging System for Artificial Organs," ASAIO J., vol. 41, no. 3, 1995, pp. M588–M595.

[36] Bove M., and Gettys J., Hardware details for OLPC, May 28, 2006, http://wiki.laptop.org/go/Hardware_specification#What_makes_this_system_unique.3F

[37] Starner T., 1996, Human-powered wearable computing, IBM Systems Journal, Vol. 35, No. 3&4, http://www.research.ibm.com/journal/sj/mit/sectione/starner.html

[38] Martinez J. U., 2002, Wireless transmission of power for sensors in context aware spaces, Master's Thesis, MIT Media Lab, p 36

[39] Sputnik, RF Propagation Basics, 2004, http://www.sputnik.com/docs/rf_propagation_basics.pdf

[40] By Choi C. Q., 2006. Miniaturized Power, Scientific American, Vol. 294, No. 2, pp 72 - 75

[41] Shaltis, P., Wood, L., Reisner, A., Asada, H., 2005, "Novel Design for a Wearable, Rapidly-Deployable, Wireless Noninvasive Triage Sensor," 2005 27th Annual International Conference of the IEEE/EMBS, Shanghai, China.

[42] iAnywhere, iAnywhere 2006, Blue SDK: An Embedded Bluetooth® Protocol Stack, www.iAnywhere.com

[43] Bahl V., 2002, ZigBee, Philips Semiconductors Division, http://bwrc.eecs.berkeley.edu/Seminars/Bahl-10.25.02/ZigBee.ppt

[44] Zombolas, C., 2003, Specific Absorption Rate (SAR) New Compliance Requirements for Telecommunications Equipment, EMC Technologies, http://www.emctech.com.au/sar/SAR_Article_2003.pdf

PART 4: MAINSTREAM WEARABLE SYSTEMS USER INTERFACES

Chapter 8

SIGHT AND SOUND USER INTERFACES

8.1 THE ROLE OF A USER INTERFACE

Of all the components of a wearable system, the one most visible to the user is the user interface. More than any other system component, the user interface can set the tone for the user's experience with the wearable system. To understand the importance of a user interface and the unique characteristics a user interface for a mainstream wearable system must possess, let's first review its role.

The principle role of a user interface is to mediate the interaction between us and the wearable system. Recall the picture of how the user and the wearable system form a larger system and the various ways they can interact. Figure 8-1 shows the overview of this system, repeated from Chapter 3.

For a PC, the vast predominance of the interface, at least today, is the Graphical User Interface (GUI). This interface composed of windows, icons, menus, and a mouse, is the interface with which we are most familiar. This type of interface is also known as the WIMP interface for Windows Icons Menus and Pointers (the mouse).

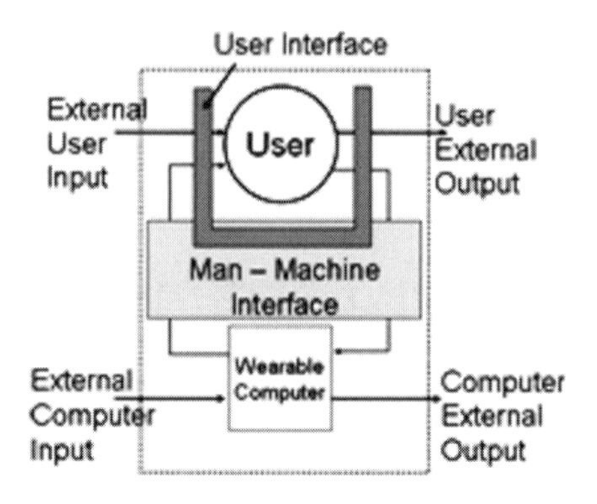

Fig. 8-1. System Formed By the User and the Wearable

For a wearable system, the scope of the user interface is much larger. Because the wearable system is mobile and personal, we can consider not only a GUI, but also speech, gesture, chording, and even eye tracking[78]. This scope expansion is driven by the need for our interface to adapt to the different situations and contexts in which we will find ourselves as we move about in our daily tasks. For instance, we may be sitting down at the table in the morning and can use the GUI, perhaps with a heads up display. Later in the day we will be walking outside where we would use a combination of speech and gesture. Still later, we will be in a meeting where we are limited to non-obvious actions such as chording.

The inclusion of these multiple types of interfaces adds complexity to the design of the system. Indeed, their inclusion is crucial if we are to allow the user easy operation of the wearable system in the many different environments the user will encounter.

[78] Chording is the simultaneous pressing of multiple keys to specify an alphanumeric input. The use of chording allows compact input devices with fewer keys. An example is the Twiddler.

There are some design principles that transcend the differences in interfaces and should be applied regardless of the type of interface mechanism (GUI, speech, etc) that is being used at the moment.

When designing any interface, we should seek to leverage learned or known cues, behaviors and organizational patterns. This reduces the amount of effort it takes to learn and use the interface.

As an example, consider the current WIMP based Graphical User Interfaces. One of the reasons they are so easy to learn is that they employ a metaphor with which most of us are very familiar, the desktop. For most of us the desktop, with its structure of folders, documents, and trashcans are very familiar since most of us work in or have exposure to the office environment. The use of a mouse is probably the most intrinsically alien element of the GUI.

Another transcendent principle of interface design is to leverage perceptible cues to increase the strength of the metaphor. In early GUIs animation and sound were used to increase the strength of the metaphor. For example, when you clicked on a folder icon, the folder icon would change to an open folder image before a window appeared showing the contents. Similarly, when you placed a document in the trash can you would hear a recorded sound of a trash can lid closing.

These cues were meant to reinforce the desktop metaphor. Such reinforcement is very important for a wearable system since the user may not be able to focus mainly on the wearable. Therefore any mechanism that can raise the signal to noise ratio of the information being presented is valuable.[79]

[79] Of course, we must be careful and not overload the user with gratuitous, non-essential feedback. As the user becomes more familiar with the system such feedback becomes less necessary and can eventually become annoying.

8.2 DESIGN ATTRIBUTES OF A USER INTERFACE

How are these goals realized in a user interface for a mainstream wearable system? There are several design attributes for such an interface. Many of them are also applicable to standard desktop GUIs. However, these attributes often have additional importance to a wearable system because of its mobile nature.

- *Intuitiveness*: This is among the most obvious of the characteristics. However, for a wearable, it is even more important. A wearable is often used in situations in which the user must absorb the information it presents very rapidly and often without full attention. Thus, even the smallest non-intuitive element can be a serious issue for a wearable interface, whereas it might be a mere annoyance in a desktop GUI.

  Besides easy to learn and understand, intuitiveness also means things act as the user expects them to. That is, the interface should react to the user's actions in keeping with the user's understanding of and experience with the world and in concert with the mental model embodied by the user interface (no POLA violations).

- *Consistent Application of the Mental Model*: No matter how intuitive the mental model of the user interface is, if this mental model is not consistently applied throughout all actions of the user interface, it will be difficult to learn. Every exception to the mental model is another piece of information the user must remember.

  Such exceptions are a major problem for the interface of a wearable system. Since the user typically divides his attention between the user interface and the real world (in which he is performing his primary task), exceptions make it very difficult to quickly and effectively complete the interaction with the wearable. The mental model embodied by the user interface must be consistently applied to its organization, navigation, behavior, and design.

- *Transparency and Clear Causality*: As user interfaces become more complex and more dependent on semi or mostly autonomous software agents, it is important that the user always understand what is happening and what will happen as a result of an action he specifies through the interface, especially for those actions that are difficult to undo or could have serious consequences. This requires that the user interface provide feedback on the progress of its actions that may take some time.

In addition, it should be able to explain why it took the action it did in response to a command the user specified. Recall that one of the basic characteristics of a wearable system is that the user is always in complete control. These capabilities will allow the user to feel 'grounded'; understanding and in control of the actions of the wearable system.

However, a balance must be struck between keeping the user informed and becoming intrusive. The amount of feedback provided can be controlled by user preferences, past knowledge gained by the learning element of the wearable, and, perhaps most importantly, by providing the information in the most efficient, context specific, and least intrusive way possible, that is, maximizing output information density[80].

- *Recover Gracefully From Errors*: Regardless of how intuitive or easy to use the user interface is, the user will make errors. It is important that the interface either provide a way for the user to easily correct the error, or the wearable itself gracefully recover from the error. If the wearable is able to recover from the error, the user interface must provide the information of that recovery and the current state of the system to the user that is relevant to the performance of the current task. This way the user will always know the current state of the system.

  In a wearable system, the user can make errors simply because he cannot concentrate fully on the wearable. He must refer to the wearable's user interface quickly so as to minimize the intrusion of the wearable on his current task. This quick referencing can increase the probability of making an error.

[80] The maximization of output information density is a mainstream wearables design principle and was discussed in Chapter 5.

Most of us are familiar with the Graphical User Interface. We use one every time we interact with a PC, PDA, or even a new smartphone. Utilizing a GUI is almost second nature to anyone who uses any of those devices regularly. In fact, the basic concept and principles of a GUI are so familiar to some people that, even though the detailed operation of a PC and smartphone are quite different, users are often able to switch from one to the other with little difficulty.

Another user interface many people with PDAs are becoming familiar with is a pen/stylus interface. Many PDAs have touch screens. In addition, tablet PCs also have touch screens. Most of these pen interfaces are based on the Windows GUI, the pen acting simply as the mouse. However, in the Tablet PC interface, Microsoft has made an effort to integrate digital ink capability into many of its applications. Some third party applications have added digital ink support as well. With digital ink, you can use the pen to make handwritten annotations within ink enabled applications, create ink based notes, and use ink for free form drawing.

Wearable devices will employ these and other interface mechanisms. We will discuss these other interfaces and revisit issues of the GUI and pen interfaces, all within the context of the use of a mainstream wearable system.

8.3 SPEECH USER INTERFACE[81]

One of the most compelling user interface mechanisms for mainstream wearable systems is the Speech User Interface (SUI). There is no denying the attractiveness of a SUI for a wearable system. Speech is the most natural method of communication. To the extent that you can address the wearable system the way you normally address people, the system is much easier to

[81] This chapter does not give a detailed discussion of how speech technologies work. There are several good references for this [1][2], [3]). Rather it concentrates on the design issues of speech interfaces for mainstream wearable systems.

use. A robust speech interface would alleviate much of the need to handle a device in order to use it. You can leave it on your body. Since a speech interface can be eyes-free and hands-free, you can engage in a dialog with your wearable to complete a task while actively engaged in another task that does require your hands and/or eyes. To appreciate the power of this capability, think about talking on the phone while preparing dinner or tying your shoes. Finally, a SUI allows the wearable to assist you in your primary task with minimal disruption to the task's performance.

There are several levels of usage of a SUI in a wearable system. At the simplest level, the SUI merely provides an alternate mechanism for entering GUI commands. This is often called "Command and Control" (C&C). With a C&C SUI, the speech interface recognizes the verbal equivalent of mouse commands or a string of mouse commands. Other commands may be recognized, including changing applications and turning the recognizer off.

The next level is to use a SUI to augment the capabilities of a gesture interface. In this case, the user gives commands that have no gesture analog. These commands however, are used to augment a gesture command. For example, the user might say "how far is this" while pointing to an object on the screen and then moving his hand to another area of the screen and saying "from that?" Often, the simultaneous use of speech with another interface mechanism will reduce the error rate of the task as each interface mechanism compensates to some extent for the uncertainty of the results in the other [4].

The final level of usage of a SUI is speech as the primary interface. At this level, most, if not all of the capabilities of the device are controlled by speech. Other interfaces such as a GUI or pen, to the extent that they are used at all, serve mainly to recover from serious user or device errors, to view information that cannot be rendered by speech such as images and video, or in situations where a speech interface cannot be used or where its use may be inappropriate.

Because of the current limitations of Automatic Speech Recognition (ASR) and Speech Synthesis (usually referred to as Text To Speech - TTS), using a SUI as the primary user interface is not viable today. However, this remains the holy grail of wearable user interface design for many.

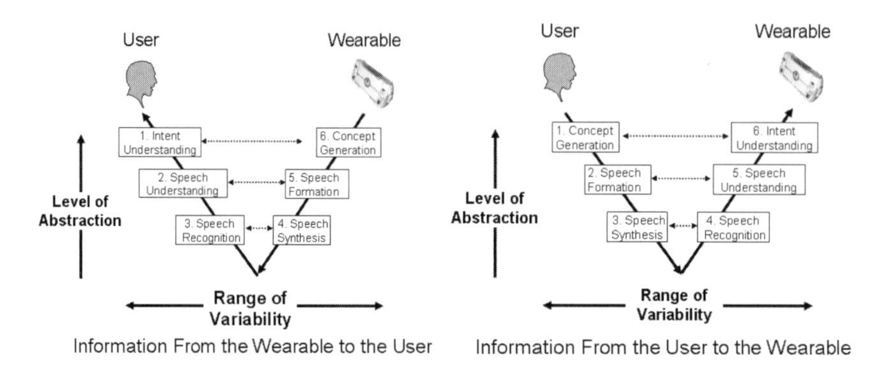

Fig. 8-2. Full Functions of a Speech User Interface

Creating a SUI that can function as the primary user interface for a wearable system involves much more than ASR and TTS. To understand the complexity of such an interface, let's look at the process of the wearable conveying a piece of information to the user and what the wearable must do to recognize and understand information from the user. We will use Figure 8-2 for this discussion.

1: Concept Generation

The first step in using speech is to conceive of a thought to be transmitted. This is a highly abstract process in humans. However, for a wearable system it would consist of invoking a specific algorithm that resulted in the need to send a verbal message to the user. This algorithm can be purely deterministic or can involve commonsense reasoning and other 'fuzzy' processes.

As an example, consider a device that received an email message. Upon receiving the message, it would determine the sender and the message's priority. It would then determine the user's current operational context. This context would include the time of day, what the user was doing (talking to another person, listening to another message, etc.), and the user's location. These elements would be evaluated against a set of notification rules the user had set or which the system had learned from the user's behavior over time.. Examples of these rules are:

- If it is between 5 pm and 7 pm, notify me only if the message is urgent or from my family.
- If I am currently talking to someone, notify me only if the message is urgent
- If the message sender is on my avoid list, delete the message without notifying me.

Note that this is basically the same process people go through when they want to tell somebody something. However, instead of the rules described above, you would unconsciously apply your preferences, social conventions, and personal notions of what constitutes polite behavior.

2: Phrasing It Accurately and Effectively: Speech Formation

Once the message is conceived, it must be put into language suitable for verbal rendering. This is referred to as speech formation. It requires knowledge of the grammar of the language used and common idioms, something humans acquire over several years of growing up within their society. These aspects of speech generation can have a significant effect on the naturalness of a speech interface.

Continuing with our example above, the device determines that it has been a while since it last spoke to you so it needs to be sure to get your attention. Therefore, it will preface the notification with your name, a common social practice. It composes the notification using the sender's name:

Joe, you have an ***important*** *message from Dr. Johnson*

It also makes sure that the article "an" is used before the word important, which begins with a vowel, and it will emphasize the word important.

3: Saying It Clearly: Speech Generation

After determining the phrase to speak, the wearable must render it as speech. This is referred to as speech generation. A wearable device would use speech synthesis. Speech synthesis has two metrics of performance: naturalness and intelligibility. Naturalness measures how much the synthesized speech sounds like a human. Intelligibility measures how easy it is to detect the words being synthesized. It is possible for a synthesized phrase to be intelligible but sound unnatural and robotic. It is also possible that the speech sounds very natural, but is hard to detect the words due to poor pronunciation rules or poor pacing of the parts of the phrase as it is synthesized.

In the example, the speech synthesizer will pause after rendering your name to allow you to change the focus of your attention. It will also emphasize the word "important" by stress and intonation. It will recognize the abbreviation "Dr." as "doctor". Finally, it will use a neutral overall intonation profile since this is a command and not a query or exclamation.

"joe   you have an important message from doctor johnson"

4: Recognizing What Is Said: Speech Recognition

When receiving speech input, the first task of the interface is to recognize what is said. This means detecting the individual word boundaries, converting each word energy input into a vector of statistics describing the word, detecting when the speech input has ended, and rejecting sound input that is not actual speech.

Each identified word is compared to other words in the current vocabulary. Various methods are used to limit the number of words that are searched for a match. Examples of search limiting mechanisms are grammars and statistical models of speech.

In our example, the user hears the device's message. He is just finishing up a task and wants to hear the message when he is finished. He prefaces each command to the device with the name he has given to the device

(Companion in this example) to indicate that he is addressing it and not someone or something else:

"Companion, read it to me in 5 minutes"

The continuous speech recognizer in the wearable system receives the input and detects the keyword "Companion". It therefore accepts the rest of the input and analyzes it to detect word boundaries. It uses its grammar for the current operational context to select the sentence that has the highest probability of matching the verbal input. In doing so, it distinguishes between the homonyms "read" and reed" and between "two", "too", and "to". It passes the selected sentence to the next element, speech understanding.

Read it to me in five minutes

5: Understanding What Is Said: Speech Understanding

Once a speech interface recognizes the words and sentences spoken, it must move from simple syntax to the more complicated element of semantics. That is, it must *understand* what was said. This means resolving the inherent ambiguities in a person's speech, including resolving pronoun references, and word senses [5]. This requires a knowledge of how words can be related in a sentence and determining the proper meaning of a word based upon where it is in the sentence.

In the example, the device uses its knowledge of the operational context to associate the pronoun "it" with the email message received from Dr. Johnson and associates "me" with the user. It then decides that "in five minutes" means a point of time in the future, as opposed to some location called "five minutes". Finally, it determines that the input is a command and not a query. And since the current operational context was email, the recognizer sends the semantic information to the email application.

6: Understanding What Is Meant – Intent Understanding

Understanding what a person really means just from a transcription of what they said is often a difficult task because speech alone does not capture all of the information available when one person interacts with another. Besides the speech itself, there are visual cues such as facial expressions and body language. In addition, with today's speech recognizers, not all of the information present in the speech is made available to the intent understanding system. For example, intonation and the pacing of the speech are not captured. This further increases the difficulty of determining what the user really means.

Finally, there is a context that underlies what a person says. This context is composed of world knowledge, past dialogs on the current task, and personal experiences. Much of this information may not be known to the device trying to discern the actual intent of the command.

The best way to deal with this uncertainty is to severely limit the context in which the speech interface operates. The less contextual information the interface must know, the more accurate it will be.

In the example, the context is limited to scheduling reminders. The email application decides that what you really want is to have the message read to you five minutes from now regardless of what you are doing unless you indicate otherwise in the meantime or it violates preset notification rules. So the device schedules a task with very high priority to read the message in five minutes.

In Figure 8-2 the actions on each side of the V connected by dotted arrows are at roughly the same level of abstraction. For example, Concept Generation on the left side of a V and Intent Understanding on the right side of the same V are roughly at the same level of abstraction. However, the range of variability, that is the potential number of errors and mistaken interpretations increases as we travel from one end to the other in the V.

Speech recognition and speech synthesis are close in the sense that there are no intervening processes and thus a few sources of errors. However, there are many intervening processes between Concept Generation by the sender and Intent Understanding of that by the receiver. Thus, as we travel from Concept Generation down the left arm of the V on the left of Figure 8-

2 and up its right arm, we traverse four intervening processes before we reach intent understanding. At each of these intervening stages errors can creep in until what Intent Understanding produces is not at all the concept originally generated. Thus the range of potential variation between what is generated on the left and what is received and produced on the right arms of the V increases as we travel down the V through decreasing levels of abstraction and then up the V through increasing levels of abstraction. The process is similar for the V on the right of Figure 8-2 illustrating information flowing in the reverse direction.

Putting It All Together: Dialog

While simple speech interfaces may involve single word or phrase input or output, more complex speech interfaces involve some element of dialog between the user and the device. Dialog is characterized by one or more bi-directional interactions between the user and the device within a specific context. The example used above presented one such interaction. In sophisticated speech applications, multiple interactions may be involved to complete a complex task. In addition to performing well at each element of speech interaction shown in Figure 8-2, the speech interface must also support the dynamics of discourse such as turn taking, barge-in, and disfluency handling. This makes conversational speech interfaces much more difficult to create.

8.3.1 Issues in the Use of Speech Interfaces

As attractive as speech interfaces are, there are several issues that must be addressed when designing them. One of the most important is to know what it means for an application to be speech enabled.

Enabling an application for speech that has been designed for GUI or keyboard/keypad input does not mean simply adding a speech recognizer and speech synthesizer. Speech, as we saw earlier, has characteristics different from those of other input mediums. This means that most applications that are designed for GUI or text/pen mediums should be redesigned to effectively utilize a speech interface.[82]

This is especially true for applications that support less structured input, such as graphics programs and programs manipulating images. In these cases, simple verbal renderings of mouse travel and selection are not effective.

Designing a speech enabled program for a wearable is a more challenging task than developing it for the desktop. The user is mobile and may not have access to a keyboard of similar input device. The user cannot devote their full attention to the speech interface since they are usually mobile and must attend to the changing environment around them.

In addition, the accuracy of the speech recognition can be impaired due to the changes in the speaker's speech. Causes for this include exertion with the current task, such as walking rapidly, differences in prosody due to emotional state, and straining to speak above the ambient noise level.[83]

A speech interface is most useful in situations when other input mechanisms are not appropriate or cannot be used. For example, when the

[82] This is true even for Command and Control speech interfaces where the user is simply giving verbal renditions of mouse commands (ex. "File" "Open", etc). For this case there will be portions of the interaction, such as selecting elements on a screen, for which verbal renditions of mouse actions are not effective. These parts of the application should be redesigned from the ground up to use speech.

[83] It is well known that a person will increase their overall vocal intensity (pitch, formant location and bandwidth, etc) in the presence of ambient noise [6]. This is the Lombard Effect.

user's hands and eyes are occupied (driving, carrying things, controlling surgical robots), their use is not practical (wearing thick gloves, displays washed out in bright sunlight), or the alternative interfaces are worse (small buttons, small displays).

However, despite their usefulness, there are situations in which a speech interface is inappropriate (in a meeting, library, church, a loud party or concert[84]) or dangerous (when stealth is necessary), or when there is no margin for error or misunderstanding (weapon fire control[85]).

Other challenges in designing effective speech interfaces for wearable systems are

- Speech asymmetry: people can speak much faster than they can comprehend while listening.
- Speech is transient so the context in which the dialog is taking place is not available to the user in real time
- Much of the semantic meaning of speech is in prosody - characteristics such as pitch contours, syllable stress, pacing, etc. Currently, speech recognizers do not recognize prosody and speech synthesizers do not render it well.
- People can speak faster than they can type, but listen more slowly than they can read
- Speech tends to be much more informal and imprecise that the written word so users are more likely to use idioms and colloquialisms

[84] The 'cocktail party effect' refers to the ability to focus listening attention on a single speaker among several others speaking at the same time. While this is relatively easy for people to do, it is very difficult for speech recognition systems [7].

[85] In this case minimizing both false positives and false negatives is important!

- A person's speech is usually not uniform. There are disfluencies such as pauses, filler words ("um", "uh", etc.), and unfinished utterances
- There can be feedback from TTS output picked up by the speech recognizer, especially if the output is directed out toward the environment as might be the case in some language translation applications
- People will often vary the order of the words for same command

Theses challenges all increase the Operational Inertia of the speech interface. We discuss possible solutions to some of these in the next section.

8.3.2 Towards Transparent Speech Interfaces

As attractive as speech user interfaces are, they currently generate significant Operational Inertia. If they are to be an integral element of the user interface of a mainstream wearable system, we must find ways to minimize or eliminate their OI[86].

One area of setup effort for speech recognizers is creating the grammars used to recognized utterances. These are usually created beforehand, compiled, and loaded into the recognizer when it is started. However, if we do not know what the utterances are before we use the recognizer, we can't create the grammar beforehand so recognition won't work.

This is most problematical in instances such as browsing the web using speech, where we could go to any web page and thus cannot know the utterances (say the hyperlinks in the page) beforehand. In this case we must dynamically create the grammars from the page we are at. By parsing the HTML and extracting the hyperlink text we can create the grammar in real

[86] Some of the approaches to reducing OI given in this and the following sections are based on emerging technologies not yet widely available. The goal is to give examples that will illustrate the application of the principles and to start a dialog of ideas on how to apply these principles to achieve transparent use design.

time, compile it, and load it into the recognizer. Then we can speak the hyperlink text and go to the associated target [8].

Another approach is to use verbal proxies. This is useful in form based input. If the field name (for example, 'Sender' in an email) is known, we can use it to indicate where in the grammar to add the field's current value. The value is added as a terminal for the associated grammar rule and now the recognizer will recognize that word once the updated grammar is compiled and loaded into the recognizer [9].

Related to this is using a virtual speech interface to control devices by speech even though they do not have a speech recognizer. We create a text file with a grammar that specifies the commands we will use in speech along with their conversion to the internal command form for the device. The wearable system then retrieves the file from the device when we want to control it by speech. The grammar is compiled and loaded into the recognizer. We can then speak the command specified by the grammar and send to the device the internal form of that command expected by the device [10].

One of the main sources of interaction complexity with a SUI is the less than perfect accuracy of the speech recognizer. To keep the vocabulary small, and thus increase the recognizer's accuracy, separate commands into their respective contexts and create separate grammars fort each context. Then automatically switch in grammars based on user context and commands.

Enforce a system wide common metagrammar which defines the structure of all commands (but not the commands themselves) used by the various speech enabled devices in the system. This reduces interaction complexity since the user must learn only one command format for the entire system. Metagrammars were discussed in 5.1.2.

Obtaining help with a SUI can generate a lot of interaction complexity. Providing help using examples instead of detailed explanations often can enhance the user's comprehension and memory of the help content and thus lower interaction complexity of putting the information to use. Using examples for help was discussed in 5.1.2.

Listener fatigue is a significant source of interaction complexity for SUIs with TTS. Summarize information and 'chunk' the content into speech output while allowing the user to barge in and stop the output at any time. Maximize output information density to minimize the amount of speech to which the user must listen.

A related issue is how to ensure the SUI gets the user's attention for output. We must strike a balance between effective notification and listener fatigue. One approach is intelligent user notification and specification, for example, speaking the user's name before a verbal message. The use of the user's name can be based upon the severity of the message and the length of time since the system last communicated with the user.

A significant source of non-use obtrusiveness for SUIs is the recognizer taking background speech not directed at it and attempt to act upon it. To prevent this, the user can explicitly activate the speech recognizer via a control (button, etc) or other user action. Alternatively keyword spotting can be used, although the rate of false positives must be very low for this method to be viable.

Table 8-1 summarizes some of the sources of the OI in speech interfaces and approaches for dealing with it discussed above.

Table 8-1. Transparent Use Approaches for Speech Interface Design

Operational Inertia Component	Sources of Operational Inertia	Design Approaches for Transparent Use
Setup Effort	• Incorporating new words, phrases	• Automatically generate grammars based on newly encountered content such as web pages
		• Use verbal proxies in which a word acts as a place holder and the actual word is inserted into the grammar from the information content
	• Accommodating devices without speech interfaces	• Devices provide a virtual speech interface textual grammar file which the

Operational Inertia Component	Sources of Operational Inertia	Design Approaches for Transparent Use
		wearable system retrieves and uses to allow the user to give speech commands to control the device
Interaction Complexity	• Insufficient recognizer accuracy	• Limit vocabulary size by automatically switching in grammars based on user context and commands
	• Different command sets among devices	• Enforce a system wide common metagrammar which defines the structure of all grammar elements (but not the phrases themselves) used by the various speech interfaces in the system
	• Listener fatigue	• Summarize information and 'chunk' the content into speech output while allowing the user to barge in and stop the output at any time
		• Maximize output information density
	• Providing effective help	• Provide help to the user with examples instead of detailed explanations
	• Getting and maintaining the user's attention	• Intelligent user notification and specification, for example, speaking the user's name before a verbal message. The use of the user's name can be based upon the severity of the message and the length of time since the system last communicated with the user
Non-use obtrusiveness	• Attempting to recognize speech not directed at the speech recognizer	• Explicitly activate the speech recognizer via a control (button, etc) or other user action

8.4 AUDIO INTERFACE

An audio interface (as opposed to a speech interface) uses non-speech sounds to convey information. The scope and ability for an audio interface to stand on its own as the primary interface for a wearable system is severely limited. It is most often employed as a supplement or complement to another interface, often speech.

8.4.1 Examples of Audio Interfaces

Audio interfaces are more useful than one may think. In 1979 strange events were plaguing the Voyager-2 spacecraft as the craft began its traversal of the rings of Saturn. Mission controllers could not determine the problem from visual displays they were receiving due to the noise in the data. However, when the data was played through a music synthesizer a "machine-gunning" sound could be heard in the few seconds where the spacecraft had been in a region of dust concentration. This helped the mission controllers determine that the problems were caused by high-speed collisions with electromagnetically charged micro-meteoroids [11].

Audio interfaces have been used extensively in seismic data analysis. Seismic data sets are very large and may stretch over many hours or days of recording. Such large data sets are difficult to analyze visually to detect small but significant event or features.

In one case analysts sped up seismic recordings 100–1600 times, to shift the low frequencies of the slow vibrations in rock to the range of human hearing. This allows many hours of data to be heard in just a few minutes, and listeners learned to discriminate nuclear bomb blasts from earthquakes with an accuracy of 90% [14].

AudioStreamer [15] presents three speech sources arranged around the user's head. One source is directly in front of the user while the other two are offset from this one by 60° to the right and left. AudioStreamer exploits a person's ability to distinguish among spatially separated voices and concentrate on one of them, a limited version of the 'cocktail party effect'.

The user indicates which audio source he is interested in by turning his head toward it. The volume of the indicated audio stream is increased, further distinguishing it from the other two. If the user looks away, the volume slowly decays to the volume of the others. If the user looks at it again, its volume is again increased.

The AudioStreamer interface also used a 400 Hz, 100 msec tone to alert the user of important information such as story boundaries in either of the two audio sources in the background while listening to the source in the foreground.

8.4.2 Issues in Audio Interface Design

An audio interface can employ several elements including

- Earcons[87]
- Audio icons
- Sonification
- Spatial discrimination

Earcons are short segments of musical tones [16]. They were originally developed to provide audio feedback of GUI actions. The audio of an earcon can vary in rhythm, pitch, timbre, register, and dynamics. Their design advantages are [12]:

- ease of production: earcons are relatively easy to construct and can be produced on a computer with tools that already exist for music and audio manipulation;

[87] Some authors include audio icons and earcons under sonification. We choose to separate them since the production and perception issues are different

- abstract representation: earcon sounds do not have to correspond to the objects they represent, so they can represent objects that either make no sound or make an unpleasant sound

However, since they are abstract there may be no natural association of the sound and the object they represent. This can make earcons harder to learn and remember, especially if a user must remember a large number of them.

Like earcons, audio icons were developed to provide feedback from GUI actions and events. Audio icons map objects and events in the interface onto everyday sounds that are reminiscent or conceptually related to the objects and events they represent [17]. For example, moving a file to the trashcan would produce the sound of a real trash can closing. Audio icons have the following advantages [12]:

- familiarity: everyday sounds are already familiar and may be understood very quickly;
- directness: everyday sounds can allow direct comparisons of length or size or other quantities.

This close association of the icon's sound with the action or event it represents means that the user can probably remember more of them than the more abstract earcons. Audio icons were used successfully in a simulation of a soft drink factory [18]. Subjects in the multiprocessing, collaborative environment quickly learned and remembered the meanings and functions of the audio icons. And Lucas [19] showed that subjects associated audio icons with their respective objects or events more easily that with earcons.

However, because of the association of real life sounds, it may not be possible to find good audio icons for some of the more abstract GUI actions such as selecting an item with the mouse.

Sonification is used to map data into sounds. Typically, an audio parameter such as duration, pitch, loudness, position, brightness, etc. is mapped to data dimension [12] Different variables can be mapped to different parameters at the same time to produce a complex sound. This approach has the following advantages:

- ease of production – existing tools allow mappings to many audio parameters;
- multivariate representation – multiple data dimensions can be listened to at the same time.

There are some issues with sonification. Some of the sounds produced can be unpleasant to listen to. This can cause user fatigue in long analysis sessions. Perceptual interactions between parameters can obscure data relations and confuse the listener.

In addition, if the dataset undergoing sonification is not dense enough, there may not be enough data points to make a sound of sufficient smoothness or duration to be readily perceived and its semantics understood. And even for long sequences or dense datasets, the resulting sounds in their aggregate may sound very unnatural or even appear to violate the laws of time or physics since their source may not correspond to a physical process [12].

Beyond these issues there is the problem of a lack of standardization of most audio interface elements. Some audio icons and earcons are standardized. Sonification mechanisms are not. This may not be as big a problem for wearables however since wearable devices are worn and usually used by only one person. If the person can configure their own audio interface elements, standardization really will not matter.

It is important to understand the limitations of an audio interface. The temptation to use earcons and audio icons for every function and action of a GUI should be resisted. Such an audio interface will be a confusing cacophony of sounds and quickly become irritating to the user.

An example of this is a study reported in [13]. A browser was augmented with an audio interface. The interface included sounds for the following HTML elements:

- headings
- hyperlinks, including previously traversed links
- ordered and unordered lists
- inline images and image maps

- addresses
- paragraph boundaries
- bolded text

While the subjects generally liked the audio demarcation of macro transitions like headings and images, they disliked the use of audio to indicate micro oriented transitions like bolded text and list elements.

8.4.3 Towards Transparent Audio Interface

We have not had a lot of experience with audio interfaces so we certainly do not know all of the sources of OI in them.

Configuring the audio interface properties can be a source of setup effort. An example of reducing this effort is to condition rendering volume on ambient noise to ensure the user hears it well, as opposed to having the user constantly adjust the volume.

Interaction complexity in an audio interface is closely related to the user's ability to understand the audio element semantics. Solutions include using audio that maps well to user's experience and world knowledge, or that evokes a semantic relationship with the event.

Because audio can be limited in its discrimination affordances, it should be used judiciously, keeping the number of audio UI elements small for easy recognition.

This also means that the use of simple audio elements for out of context (i.e. not related to the user's current primary task) notifications should be minimized. They are a source of non-use obtrusiveness of the audio interface.

Table 8-2 summarizes sources of OI with some possible approaches discussed above to minimize or eliminate them.

Table 8-2. Transparent Audio Interface Design Approaches

Operational Inertia Component	Sources of Operational Inertia	Design Approaches for Transparent Use
Setup Effort	• Configuring the audio rendering	• Condition rendering volume on ambient noise to ensure user hears it well
Interaction Complexity	• Understanding audio element semantics	• Use audio that maps well to user's experience and world knowledge • Use audio that evokes a semantic relationship with the event • Do not overuse. Keep number of audio UI elements small for easy recognition
Non-use obtrusiveness	• Output not associated with primary task	• Minimize simple audio elements for out of context notifications.

8.5 GRAPHICAL USER INTERFACE FOR WEARABLES

Wearables have a love-hate relationship with the Graphical User Interfaces (GUI) built around the WIMP paradigm. On the one hand, the GUI is, in general, very ill suited for a wearable system. On the other hand, it can be the most effective interface for a wearable system for certain tasks.

There is no doubt that a GUI[88] is ill suited as a general interface for mainstream wearable systems. There have been several studies and personal

[88] In this section, the term GUI refers to a GUI employing the WIMP paradigm.

anecdotes of the difficulty using a GUI with a wearable computer [20], [21] [22],[23].

Rhodes in [20] details an experience in which he searched for a specific street via his wearable computer. He found the WIMP based interface to his wearable computer was highly ineffective for this task. A similar experience is reported in [21] where the author attempted to use a Poma, a consumer oriented wearable computer while walking down the streets of New York City. The main reason for these difficulties is that a WIMP interface makes several assumptions about the user, the computing environment, and the devices:

- The user has fine motor control. Or, more accurately, the user can apply fine motor skills to the current task. The mouse is a high precision pointing device, typically with pixel level resolution. The small size of the display (for example, a heads up display) increases the degree of fine motor skill required.

  To be effective, its use requires a stable platform upon which to move. In most cases we will not have a stable platform when using a wearable system. The hand holding the pointing device will not be resting on anything. This requires that we rely on our arm muscles to provide a stabilization force. In addition, we will frequently be moving while trying to use the pointing device. The body movements while walking introduces instabilities into the arm, increasing the difficulty of performing fine motions.

- The user has screen real-estate to burn. The WIMP interface assumes a desktop monitor. These displays are relatively large – some recent displays are 27 inches diagonally or more. Even at these sizes display clutter can be a problem as multiple windows, icons, and toolbars compete for space. Trying to present all of that visual information in the tiny screen of a heads up display usually results in an interface that can overwhelm the user as they try to discern very small elements of windows (close and minimize boxes) and icons.
- Dealing with the computer is our primary task. As mentioned in the discussion of Operational Inertia in Chapter 4, dealing with our electronic devices is usually not our primary task. In a mobile environment we must attend to the environment around us and to our primary task. In the WIMP interface precise mouse movement, selection

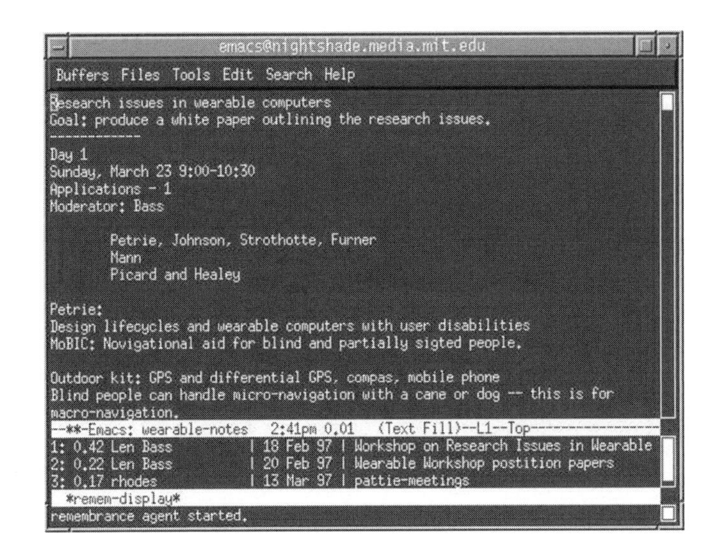

Fig. 8-3. Remembrance Agent Screen Shot *(MIT Media Lab/Alex Pentland)*

of small elements such as window close boxes and individual text characters all require a level of concentration we cannot normally afford when using a wearable system.

Figure 8-3 shows a screen shot of the Remembrance Agent (RA) display [24]. RA is an application developed for a wearable computer that manages user's information. The display requires significant user focus given the high density of information.

At a minimum, a GUI interface for a wearable system must have a display and a pointing and selection device to provide the functions of the desktop mouse. Other devices, such as a text input device may be highly desirable, but are not absolutely necessary to provide a GUI for a wearable system.

The most demanding of the two, both from a technical and a human factors point of view, is the display. We confine our discussion to displays that can be worn while being used. These include:

- Displays sown into garments. These consist of both fabric and non-fabric based display material. Fabric based materials include thin electroluminescent substrate woven between row and column electrodes developed by Visson Ltd. [25].

Transversal Light Emitting Optical Fibers .25mm in diameter can also be woven in weft positions on 2D looms [25]. The basic fabric provides fixed patterns or logos that can be lighted. More intricately woven fabrics provide thinner patterns that allow rows and columns to be independently illuminated for dynamic matrix displays.

Non-fabric based materials include Organic Light Emitting Diodes (OLED) and electronic ink. These materials use flexible substrates. OLEDs are attractive because they do not require a backlight. This means that they draw significantly less power than backlit LED displays.

Electronic ink displays are attractive for wearable systems because they have high contrast and reflect light well. This makes them easy to see in daylight. They are also capable of maintaining the text and images indefinitely without drawing power. Power is only required when the display contents are changed.

- Displays carried, but held to use. These displays tend to be about 7" diagonal and are carried in a case the user wears. Many now have a wireless connection to the wearable system's central unit. These devices require two hands to operate and are usually pen driven. Because of their size and weight they have a lot of non-use obtrusiveness when being worn and can have significant in-use obtrusiveness (part of interaction complexity) when held for use.
- Head Worn Display (HWD)[89]. Common approaches include embedding a HWD in a pair of glasses [26], attaching virtual display units to the glasses frame [27], and providing a dedicated head worn frame which holds the display units [28].

[89] Many people refer to these displays as Heads Up Displays (HUDs). However, we use the term Head Worn Display (also called Head Mounted Display (HMD)) to distinguish them from HUDs that are not worn, such as systems that project combat information onto a glass in the front of a pilot's cockpit.

These displays are either totally immersive or look through. Totally immersive HWD do not allow the user to see the environment and present a solid barrier to the eyes.

There are many candidates for the pointing and selection device including:

- Chording devices such as the Twiddler, a device held in one hand with several buttons [29]. The user uses single and multiple, simultaneous button presses to enter information. It also contains the IBM Trackpoint as the mouse and uses buttons as the selection buttons. Chording devices can be difficult to learn or uncomfortable to operate. However, Starner has reported good success in both quickly teaching people how to use the Twiddler and in their typing speed using the device [31].
- Keyboards worn on the body, typically the wrist [30]. These tend to be either rigid full or half QWERTY keyboards, or flexible keypads sown into clothing. The rigid keyboards are very obtrusive, often taking up most of the forearm. Using a wrist keyboard makes data entry a two handed process. In addition, if the data entry requires an extended amount of time, the effort required to keep arm with the keyboard in its position (arm parallel to the ground and bent inward at the elbow) can cause fatigue, making continued entry difficult.
- On screen keyboards. These 'soft' keyboards appear in the display and the user selects each character with a selection device. A wearable system user could use a Twiddler or similar device to select the letters appearing in a wearable display. Soft keyboards can be difficult to use, since fitting an entire QWERTY keyboard into the small displays of a wearable system results in very small key cells.

GUIs are far from an ideal user interface for a mainstream wearable system. However, the reality is that many if not most of the systems will incorporate a GUI as one of its user interface mechanisms. The question thus becomes: how can we make a GUI for a wearable system transparent? This is the focus of the next section.

8.5.1 Towards Transparent Graphical User Interfaces

The current use of obtrusive head worn or body worn displays and obtrusive input devices, both of which require close and sustained user focus, makes a GUI utilizing the WIMP paradigm something of an oxymoron for a transparent wearable system. Nevertheless, let's consider the basic design requirements for a GUI suitable for use with a transparent wearable system.

For a GUI UI, setup effort consists of putting on the display and input devices and orienting and configuring them for use. To minimize display setup effort, incorporate the display into something the user already wears such as eyeglasses. Issues of display orientation can be minimized by designing devices to sense their current orientation and adapt the presentation of the information to it. And for devices such as displays that are not embedded, a swing/flip down into viewing position configuration can reduce setup effort, at a possible cost of added obtrusiveness.

There are several sources of interaction complexity with a GUI. Selecting objects is one source. Display information such that it does not require high physical precision positioning for selection. For example, if using a speech with a GUI in a command and control interface, design display elements with short, polysyllable tags with minimal confusability with other tags.

How information is displayed in a GUI UI can also be a source of considerable interaction complexity. To ensure that the display provides information that can be comprehended at a glance, use high contrast, sparse, concise text with minimal graphics. When overlaying information on real world view, such as in Augmented Reality applications, overlay most information at margins of display; minimize overlay clutter. Minimize detail required to comprehend display, that is, maximize Output Information Density. And adjust display brightness to ambient light to minimize glare and excessive display brightness that can cause eye fatigue

Another significant area of interaction complexity for a GUI is inputting information. The system can employ context to present targeted, highly probable choices of inputs to user, minimizing search and selection time. When there is only a single input relevant to the current task and it is known to the system, that is, the input space is fully constrained, the system should input the information for the user. The display should show the insertion, allowing the user to abort or back out of the decision taken by the wearable system.

One of the most challenging design tasks is to provide input affordances that fully accommodate size of the user's fingers, while minimizing the size of the device to reduce obtrusiveness. Ideally, input devices such as wearable keyboards and Twiddlers would be sold in multiple versions, each one containing key sets of a different size so users could select the size they wanted and make their own tradeoff of interaction complexity (using the device) and non-use obtrusiveness.

Non-use obtrusiveness of a GUI interface is made up of the non-use obtrusiveness of the GUIs physical affordances – displays and input devices, and the non-use obtrusiveness of the output it provides.

Obtrusiveness of body worn displays can be minimized by incorporating them into NZOID devices such as eyeglasses. In addition, using flexible displays that conform to the body's contours will also minimize their obtrusiveness.

Non-use obtrusiveness of input devices can be minimized by ensuring the devices have no protruding, concave, or sharp outer surfaces and that they are attached to the body in a way that accommodates user motion and changes in posture.

Unsolicited user output can also increase non-use obtrusiveness. Minimize display real estate used for the output to minimize the time the user must spend to absorb the unsolicited output; again, maximize Output Information Density.

Table 8-5 summarizes sources of OI with the approaches discussed above to minimize or eliminate them.

Table 8-5. Design Issues for Transparent WIMP GUI

Operational Inertia Component	Sources of Operational Inertia	Design Approaches for Transparent Use

Operational Inertia Component	Sources of Operational Inertia	Design Approaches for Transparent Use
Setup Effort	• Donning the display	• Incorporate the display into something the user already wears such as eyeglasses
	• Orienting/configuring display for use	• If not integrated into a host device, swing/flip down into viewing position
		• Design devices to sense their current orientation and adapt the presentation of the information to it
Interaction Complexity	• Selecting objects	• Do not require high physical precision positioning for selection
	• Density of information on display	• Use high contrast, sparse, concise text with minimal graphics for 'at a glance' comprehension
	• Overlaying information on real world view	• Overlay most information at margins of display; minimize overlay clutter
	• Inputting text, other information	• System completes information input in fully constrained cases
		• Use context information to present targeted, highly probable choices of inputs to user
		• Provide input affordances that fully accommodate size of user's fingers
	• Eye fatigue	• Minimize detail required to comprehend display
		• Maximize Output Information Density
		• Adjust display brightness to ambient light to minimize glare and

Operational Inertia Component	Sources of Operational Inertia	Design Approaches for Transparent Use
		excessive display brightness
Non-use obtrusiveness	• Obtrusiveness of body worn display	• Incorporate display into glasses
		• Use flexible displays that conform to the body's contours
	• Obtrusiveness of input devices	• No protruding, concave, or sharp outer surfaces
		• Attach devices to the body in a way that accommodates user motion and changes in posture
	• Unsolicited user output	• Minimize display real estate used for the output; maximize Output Information Density

The interfaces we discussed in this chapter can be found in computers today. In the next chapter we discuss some interfaces that are emerging and will likely be found in mainstream wearable systems.

REFERENCES

[1] Rabiner L., Juang B., 1993, Fundamentals of Speech Recognition, Prentice Hall PTR

[2] Tatham M, Morton K., Developments in Speech Synthesis, 2005, John Wiley & Sons

[3] Reiter E., Dale R., Bird S., et. al., 2006, Building Natural Language Generation Systems, Studies in Natural Language Processing, Cambridge University Press

[4] Kvalen K., Warakagoda N., and Knudsen J., (2003), Speech Centric Multimodal Interfaces for Mobile Communication Systems, Telektronikk, pp 105-106

[5] Inman D., (2002), Ambiguity, NLP Tutorials, available at http://www.scism.sbu.ac.uk/inmandw/tutorials/nlp/index.html, accessed March 25, 2006

[6] Chi S. and Oh Y., 1996, Lombard Effect Compensation and Noise Suppression for Noisy Lombard Speech Recognition, ICSLP 96. Proceedings., Fourth International Conference on Spoken Language

[7] Arons B., 1992, A Review of the Cocktail Party Effect, Journal of the American Voice I/O Society

[8] J. Dvorak, "Audio interface for document based information resource navigation and method therefor", U.S. Patent 5,884,266, Jul. 6, 2004

[9] J. Dvorak, "Methodology for the use of verbal proxies for dynamic vocabulary additions in speech interfaces", U.S. Patent 6,473,734, Oct. 29, 2002

[10] J. Dvorak, "Virtual speech interface system and method of using same", U.S. Patent 6,760,705, Mar. 16, 1999

[11] Kramer G, 1994, Auditory Display: Sonification, Audification and Auditory Interfaces. SFI Studies in the Sciences of Complexity, Proceedings Volume XVIII. Addison Wesley, Reading, Mass.

[12] Barrass S and, Kramer G., 1999, Using sonification, Multimedia Systems, Springer-Verlag, 7: 23–31

[13] James F., 1996, Presenting HTML Structure in Audio: User Satisfaction with Audio Hypertext, Technical Report: CS-TN-98-68, Computer Science Department, Stanford University

[14] Hayward C (1994) Listening to the Earth Sing. In: Kramer G (ed) (1994) Auditory Display: Sonification, Audification and Auditory Interfaces. SFI Studies in the Sciences of Complexity, Proceedings Volume XVIII. Addison Wesley, Reading, Mass., pp 369–404

[15] Schmandt C. and Mullins A., 1995, AudioStreamer: Exploiting Simultaneity for Listening, CHI'95 pp. 218 - 219

[16] Blattner M, Sumikawa D, Greenberg R (1989) Earcons and Icons: Their Structure and Common Design Principles. Hum Computer Interaction 4(1): 11–44. Lawrence Erlbaum Associater. Mahawah, New Jersey

[17] Gaver WW (1994) Using and Creating Auditory Icons. In: Kramer G (ed) (1994) Auditory Display: Sonification, Audification and Auditory Interfaces. SFI Studies in the Sciences of Complexity, Proceedings Volume XVIII. Addison Wesley, Reading, Mass., pp 417–446

[18] Gaver w. w., Smith R. B., and O;Shay T., 1991, Effective sounds in Complex Systems: the ARKola Simulation, Proceedings of CHI'91, pp. 85 – 90.

[19] Lucas P. A., 1994, An Evaluation of the Communicative Ability of Auditory Icons and Earcons. In: Kramer G, Smith S (eds) Proceedings of the Second International Conference on Auditory Display ICAD '94, Santa Fe, New Mexico. 7–9 Nov, 1994

[20] Rhodes, B., 1998, WIMP Interface Considered Fatal, IEEE VRAIS 98 Workshop on Interfaces for Wearable Computers

[21] Alpert M., 2002, Machine Chic, ScientificAmerican.com, http://www.sciam.com/article.cfm?articleID=0003640F-8106-1D2B-97CA809EC588EEDF&pageNumber=1&catID=2

[22] Clark A. F., 2000, What do we want from a wearable user interface?, Proceedings of Workshop on Software Engineering for Wearable and Pervasive Computing

[23] Bass L., Mann S., Siewiorek D. et. al., 1997, Issues in Wearable Computing: A CHI 97 Workshop, SIGCHI, Vol.29 No.4, October 1997

[24] Rhodes, B., The wearable remembrance agent A system for augmented memory, The Proceedings of The First International Symposium on Wearable Computers (ISWC '97), Cambridge, Mass, October 1997, pp. 123-128.

[25] Deflin E., 2001, Bright Optical Fibre Fabric, The 6th Asian Textile Conference – Hong Kong – August 22-24, 2001

[26] Lumus PD-20 Presentation, 2006, Lumus Ltd, http://www.lumus-optical.com/Downloads/Tech/LumusPD-20Series-General.pdf

[27] MicroOptical, Inc., 2006, SV-6 PC Viewer, http://www.microoptical.net/Products/vga.html

[28] Bishop T., 2006, EyeBud can turn video iPod into big-screen TV for one, http://seattlepi.nwsource.com/business/254134_ipodscreen02.html

[29] Twiddler2, Handykey Corporation, http://www.handykey.com/site/twiddler2.html

[30] WristPC Keyboard, 2006, L3 Systems, http://www.l3sys.com/keybd/keybd.html

[31] Lyons K., Starner T., Plaisted D., et. al, 2004, Twiddler Typing: One-Handed Chording Text Entry for Mobile Phones, Conference on Human Factors in Computing Systems

Chapter 9

EMERGING USER INTERFACES

9.1 GESTURE INTERFACES

Gesture interfaces have been around since the early 1920's when Russian physicist, Leon Theremin invented the instrument that bears his name. The Theremin is played by moving your hands above the instrument. Two antennas protrude from the base holding the electronics. The vertical antenna controls the pitch, and the horizontal antenna controls the volume. Moving the hand closer to the vertical antenna increases the pitch. Moving the hand closer to the horizontal antenna decreases the volume [1].

However, the general public most likely conceives of a gesture interface as the one shown in the 2002 movie "The Minority Report". In the movie Tom Cruise is shown wearing gloves with bright LEDs on the fingertips. He manipulated images on a large screen in front of him by moving his hands. His movements were scripted and were highly mnemonic of the command's effect.

A recent application of a gesture interface that invokes the spirit, if not the sophistication, of the interface in "Minority Report" is the Atlas Glove (see Figure 9-1) [2]. The user wears a glove on each hand consisting of a bright white light bulb. Standing in front of a camera connected to a PC, the user makes predefined gestures to control the display of Google Earth maps

Fig. 9-1. Atlas Glove Gesture Interface *(CC - Some Rights Reserved – Dan Phiffer & Mushon Zer-Aviv)*

using pan, zoom, rotate, and tilt. Squeezing the glove turns the bulb on. The project is open source and the implementers have released the source code (written in Java) on their web site for downloading.

Devising an effective and easy to use gesture interface can be challenging. The first task is to make sure the application or system is suitable for a gesture interface [3]. If the application requires commands that indicate spatial relationships with high accuracy, or if significant part of the command's input needs to be text, a gesture interface is probably not appropriate.

Gesture interfaces are good for applications that do not required high precision spatial positioning, those that do not require complex command structures, and whose command semantics can map easily to gestures.

Designing the Gesture Vocabulary

A crucial step in gesture interface design is designing the proper 'gesture vocabulary'. There are many different types of gestures [4]:

- Emblematic gestures typically form symbols that represent specific words; for example, the circle formed by the thumb and middle finger to represent 'OK'.[90]
- Propositional gestures indicate measures in the space around the user. They are often used to illustrate sizes or movement. An example is spreading your hands apart to indicate the size of an object.
- Iconic gestures illustrate features in events and actions, or how they are carried out. An example is mimicking the movements of an action such as typing on a wrist worn keyboard.
- Metaphoric gestures are similar to iconic gestures, but represent abstract depictions of non-physical form. An example is rotating your hand at the wrist as a sign to speed something up.
- Deictic gestures refer to the space between the user and those interacting with him. An example is pointing to a spot on the floor where someone should be standing.
- Beat gestures emphasize words. They are highly dynamic but do not concretely stand for spoken words. An example is moving your hand downward in a chopping motion to emphasize a point.

Some of these gestures are done consciously, some are done unconsciously, and others can be both. Emblematic gestures are typically done consciously while propositional, metaphoric, and beat gestures are usually done unconsciously. In addition, iconic, emblematic, and metaphoric gestures are usually culturally dependent while the others are not.

Gestures that are typically done consciously (i.e. iconic, metamorphic, and emblematic) are usually modeled in gesture interfaces. Since they are done consciously, their context and meaning can be specified. This provides a predictable, repeatable mapping between the gesture and its semantics.

[90] Emblems, because they represent specific symbols, can be culturally dependent. For example, the 'OK' symbol referenced is for Western societies. The same gestural symbol represents money in Japan [5].

Unconscious gestures are usually not modeled in gesture interface systems. However, they often are a more reliable indicator of what a person means than the words spoken.[91]

Many current gesture interfaces treat gestures as a self contained language. For example, the Gesture Pendant [6] allowed a user to control objects in the house through hand gestures. Six gestures were defined: window up, window down, fireplace on, fireplace off, door open, door close. No other interface was used. This interface was in the spirit of the gesture as standalone language interfaces in Minority Report and the Atlas Glove.

Another method of using a gesture interface is to complement a speech interface by providing spatially oriented parameters for the spoken command. As an example, the crisis management system in [7] uses gesture in this way. The user can say "… a flow direction in this way with impact in these areas" and then use a hand to outline a rectangular area on a map. The gesture interface resolves the motions to locations on the map and these are passed as parameters of the spoken command.

Using a gesture interface in this way can increase the overall performance of the multimodal user interface. The gesture interface input can help to correct errors made in the speech recognizer and vise versa.

Despite the emphasis on conscious gestures, if the gestures are to be easy for even the casual user of the interface to remember, they should be built upon those gestures we do naturally. And those that we do the most naturally are often done in concert with speaking. These gestures are usually not done consciously. Their meaning is dependent on context, that is, what the person is saying at the time the gesture is made. This suggests that a gesture

[91] A person may make a mistake and say "left" when they mean "right". However, they will probably point towards the right. Members of the audience may internally correct such spoken errors, when seeing the speaker's gestures [5].

interface may be able to compensate for a speech interface's lack of prosody recognition.

This conclusion is bolstered by the observation that that the distribution of gestures during speech is similar to that of intonation patterns [4]:

- gestures are isomorphic with intonation. For example, during speech the speaker's hands rise into space with the rise of intonation at the beginning of an utterance, and the hands fall at the end of the utterance along with the final unit of intonation
- the part of the gesture with the most emphasis (the "stroke") occurs with the pitch accent, or most forceful part of enunciation

Thus, unconscious gestures may track elements of prosody in speech. This provides a mechanism for obtaining the semantics embedded in the prosody that are not processed by the speech recognizer.

There is often the temptation to select gestures that are the easiest for the gesture recognition system to recognize. This set will vary depending on the rccognition mcthod (accclcromctcrs, vision, ctc).

However, this can lead to gestures that have no mnemonic relationship to the command semantics and can be problematic when considering the ergonomics and biomechanics of the fingers, hand, and arm. These gestures can be muscularly stressful and, in extreme cases, may be impossible for some people to do.

Instead, the gestures should be designed for minimum user effort. This means they will be:

- easy to perform and remember. For example, if using both hands for the gesture, it should not require a high degree of coordination between the two hands since this can make the gesture difficult to perform by some people;[92]

- intuitively mapped to the gesture semantics;
- ergonomic; not physically stressing when used, even by frequent users. The gestures should use relaxed muscle configurations whenever possible.

Of course, it must still be possible for the system to recognize the gestures. This means:

- each gesture must be sufficiently different from all others so there is no ambiguity and
- the more features of the arm, hand, and figures the recognizer considers the better the accuracy but the complexity of the recognition processing increases.

In addition, it is important that the recognizer capture the gesture quickly to avoid requiring the user to maintain the gesture action for an excessive amount of time.

When deciding on the gestures it is important to test them early in the design process by having naïve users perform the gestures in use case scenarios and having other naïve users attempt to determine what the gestures mean. A process for designing gesture interfaces is described in more detail in [3].

9.1.1 Towards Transparent Gesture Interface

We have not had a lot of experience with gesture interfaces so we certainly do not know all of the sources of OI in them.

[92] Everyone is familiar with trying to use one hand to pat the top of the head while the other is rubbing the stomach in a circular motion. This is a common gesture pair that most people find very difficult to do well.

Setup effort for a gesture interface involves calibrating and refining the gestures for accurate recognition. For maximum transparency, use normal user actions/gestures and the user's context to calibrate gesture sensors. For example, the gestures may be different (amplitude and trajectory) if sitting in a car vs. standing up.

Adding new gestures can be a source of setup effort if the user must explicitly perform a separate gesture learning process. Instead, the system monitors the user's gestures and uses of other UIs (speech, haptics, etc) to 'learn' the new gesture and adds it to the gesture vocabulary.

Everyone makes gestures differently and the same person can make them differently depending on the context. To increase gesture recognition accuracy and reduce interaction complexity, the system can employ adaptive, learning gesture profiles that can dynamically conform to the way the gesture is currently being done.

Because gestures are spatial, describing them for help can be difficult. The system can record gestures as they are established and render them as animation for help. This will make the gesture help much more understandable.

Like all other UIs in the wearable system, the gesture interface will operate under different environmental conditions. Chief among these for vision based gesture recognizers such as the Gesture Pendant is lighting conditions. One possible solution is to provide hand lighting which reacts to ambient light level to ensure the hands can be seen well enough for accurate gesture recognition.

Recovering from interface errors and non-recognition is a source of interaction complexity for a gesture interface as it is for all UIs. Upon error or non-recognition, supplement the gesture with another UI (speech, GUI) to increase recognition probability and accuracy. For example, say the gesture while doing it. In other cases, use a GUI to select the action from a menu of gestures in the vocabulary.

Table 9-1 summarizes some of the expected sources of OI with some possible approaches discussed above to minimize or eliminate them.

Table 9-1. Transparent Gesture Interface Design Approaches

Operational Inertia Component	Sources of Operational Inertia	Design Approaches for Transparent Use
Setup Effort	• Calibrating gesture recognition system	• Use normal user actions and user context to calibrate gesture sensors
	• Adding new gestures	• System monitors user's gestures and uses of other UIs (speech, haptics, etc) to 'learn' the new gesture and adds it to the gesture vocabulary
Interaction Complexity	• Obtaining help for gestures	• Record gestures as they are established and render them as animation for help
	• Accommodating individual movement characteristics	• Employ adaptive, learning gesture profiles
	• Varying lighting conditions for vision based systems	• Provide hand lighting which reacts to ambient light level to ensure the hands can be seen well enough for accurate gesture recognition
	• Recovering from interface errors and non recognition	• Upon error or non-recognition, supplement with another UI (speech, GUI) to increase recognition probability and accuracy.
Non-use obtrusiveness	• Obtrusiveness of camera or gesture sensors	• Tradeoff thickness for surface area in sensor design

9.2 HAPTICS INTERFACE

Haptics typically refers to sensing and manipulation through touch. A haptic interface has two components: tactile sensing, and kinesthetic sensing. Tactile is an awareness of stimulation to the outer surface of the body (for example, the vibration of a cell phone in a person's hand). Kinesthetic sensing is an awareness of limb position and movement (for example, an ability to touch your nose with your eyes closed), as well as muscle tension (for example, estimation of object weights) [8].

There are several sources of haptic sensation [14]:

- force / torque
- vibration or impulse
- motion arrest - brake
- temperature
- pressure - inflate/deflate/vibrate
- touch – make/break physical contact

Most people are familiar with haptics as an output medium. Examples include Braille readers, phone vibrators, and force feedback mechanisms in game playing controls.

Haptics is an emerging field and is being applied to wearable devices including cell phones. Moving beyond the current vibration mode for incoming calls, haptics will allow you to distinguish among a small number of callers by vibration patterns. Adding haptics to SMS messages and other non voice communication could enhance the experience. Imagine for example, feeling a heartbeat when receiving a message from a loved one. Haptics can also provide tactile feedback for soft keyboards. The user would feel a localized tactile confirmation that the soft key was pressed.

However, haptic interfaces will become more sophisticated than this. An early example of a haptics interface is the Optacon (Optical to Tactile Converter) [9]. Designed to aid blind users to read text, it is no longer made. It consisted of a small handheld camera connected to a box housing the electronics and an array of pins. The 6 x 4 pin array matched the 6 x 4 array of photocells in the camera unit. The user placed one hand on the pin array and held the camera in the other. As he swept the camera over the text, the photocells would register the dark areas on the page. This would cause the corresponding pins in the pin array to vibrate. Thus the image of the page was transferred to the user's hand via the pin array. A highly trained user could read up to 100 wpm, with typical rates of around 50 wpm. Since the camera simply transferred black areas to the pin array, the Optacon could be used to perceive images as well.

More recent examples are the Rutgers Master II [10] and the CyberGrasp system (see Figure 9-2) [11]. The CyberGrasp system consists first of a fabric glove called the CyberGlove. The CyberGlove has 22 sensors which measure the joint angles of the fingers, hand and wrist. Attached to the top of the glove is an exoskeleton which provides force feedback to the user. It guides force-applying mechanical tendons to the user's fingertips. The tension in the tendons is controlled by the actuators located in an electronics box attached to the exoskeleton via multiple cables.

Haptic input has also been used in affective computing. In [12], users were subjected to tasks that were deliberately designed to induce frustration and stress. Electromyographic (EMG) signals from seven muscles on each subject were recorded during the tasks and the mouse used by the subjects was augmented with pressure sensors on the sides and top. Incorporating this haptic input into the user interface of a wearable system could record the levels of user frustration, allowing the wearable to take action to reduce the frustration or minimize its effects. This could be especially valuable in driving situations [13].

Several haptics displays have been built into clothing [[15], [16], [17]]. Gemperle st. al [[15]] makes the distinction between a haptics device, a sensory assistive device, and a haptics display. A haptics device provides some force feedback when an event happens. An example is a haptics enabled mouse that pushes against its direction of motion when it is rolled over the edge of a window, creating the sensation of an 'edge' on the window [18]. Haptics devices take their cues directly from the visual interface.

Sensory assistive devices typically aid the deaf or blind in their ability to perceive the world around them. These devices use tactile stimulation to translate audio or visual information to touch. For example, the Opticon takes visual images and creates a texture map via the pin array. There is direct mapping from the visual image to the haptic texture map.

A haptics display can be defined as "... a device which presents information to the wearer by stimulating the perceptual nerves of the skin."

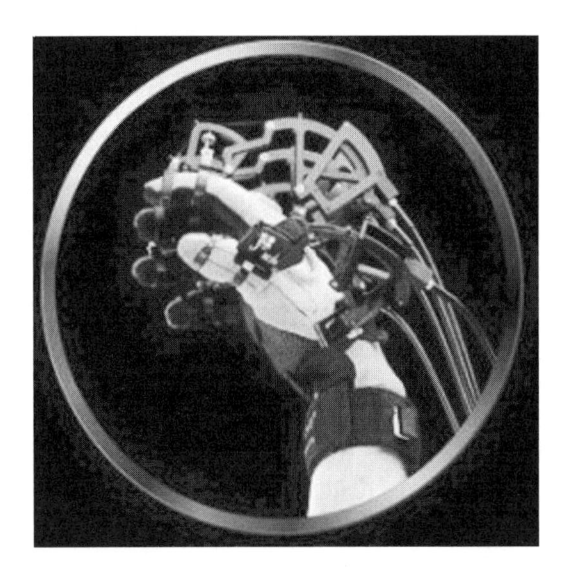

Fig. 9-2. The Cybergrasp Force Feedback Glove *(Reproduced by permission of Immersion Corporation, Copyright ©2006 Immersion Corporation. All rights reserved)*

[15] Whereas haptics devices and sensory assistive devices typically provide a direct mapping or a translation of real or computerized visual or audio information into tactile stimulation, a tactile display is neither direct nor a translation of visual or audio information to touch. Haptic displays present tactile information with its own semantics.

A wearable haptics display must meet several requirements to be effective [15][93]:

- It must be light weight, since it may be worn for some time and should not fatigue the wearer.

[93] Most of these requirements apply to vibration actuators. However, size, weight, and sound apply to all types of wearable haptic elements.

- It should operate silently. The tactors (tactile stimulators) should be muffled or otherwise conditioned to make the minimal noise possible. The input should be the tactile sensation, not sound.
- The tactors need to be very small since they will be embedded in clothing or a portable device or attached directly to the body. However, there is a tradeoff between a tactor's size and the intensity of tactile stimulation it can produce. As a rule, the smaller the size of the tactor, the less it's mass and hence the less intense its haptic effect.
- The tactor should consume as little power as possible. Incorporating a power supply into each tactor is impractical since it would significantly increase its size. Thus, all of the tactors (of which there could be many in the garment [16], [17]) must derive power from the power distribution network in the garment or device.
- The user must feel the tactors through all the garments between the tactor and the user's skin, where the sensation will register with the body. This includes the garment containing the tactors. Depending on the characteristics of garments (thickness, material, etc) between the user and the tactor enhanced garment, the tactile sensation will vary. Thus, it is possible that the user will easily perceive the tactors when wearing a thin shirt but not when wearing a sweatshirt underneath the tactor enhanced garment.
- For the user to have a chance of perceiving haptic input, the tactors must be held tight on the body. This puts serious constraints on the design of garments that would provide haptic input.
- The tactors must each be physically discreet and sufficiently separated from each other so that the user can easily tell which sensor is activated. The tactors themselves should be placed in areas of the body that convey meaningful semantics. For example, if you are providing navigation, the tactors should be placed near the right and left edges of the body [19].

  In addition to those above we must add:

- The garment must minimize the stimulation propagation so it remains localized near its source [17]. This increases the ability of the user to easily tell which tactor or which area of the body is being stimulated.
- The actuators must be incapable of harming the user, even if they malfunction. For example, if we are using sensors that provide input via

temperature, it should not be possible to burn the user even if the sensors malfunction and operate beyond their expected temperature range.

Haptics is often used with other interface mechanisms [18], [20], [21]. In [18] a shirt is augmented with a sonar transceiver and tactor on each shoulder and a microcontroller and D/A converter on the front center. The sonar transceivers use ultrasound pulses to detect the presence of objects within their field of view, about 60°. Since the system is stereoscopic, it can indicate to the user if the object is on the user's left or right by vibrating the tactor on that side. If both tactors activate, the object is directly in front of the user, or extends across both sonar transceiver's view. With proper training the user can also detect the height of an object, and can estimate the position and speed of moving objects.

The UltraCane [21] also uses sonar ranging and vibration feedback. However, the transceivers and tactors are embedded in a cane similar to the white canes used by the blind.

9.2.1 Towards Transparent Haptics Interface

We have not had a lot of experience with haptics interfaces so we certainly do not know all of the sources of OI in them.

Placing tactors on the body and connecting them to the body network is a significant source of setup effort. Embedding them in a tight fitting garment to ensure good body contact is one approach to overcoming this.

Haptics interfaces involving exoskeletons for force feedback can also involve a lot of setup effort. Placing the exoskeleton framework on a wearable item such as a glove minimizes the setup effort.

Interaction complexity for a haptics often involves how readily the user can sense and identify the individual tactor or the pattern caused by multiple tactors. To ensure good perception the tactors should be separated sufficiently to prevent vibration propagation from the activated tactor to another, non-activated tactor. However, in the case of coordinated tactors forming a pattern, the inter-tactor spacing should ensure that both the pattern as a whole and the direction of any movement is easily perceived.

Perception of tactors can be increased by placing the tactors on areas of the body that do not deform significantly with changes in user posture, for example the back and chest. However, when placing tactors, the semantics of the tactor activation should also be considered. For example, if the tactor activation signals direction, they should be placed as far from the body centerline as possible.

Vibration patterns tend to be abstract and, as such, can be hard to remember and differentiate in large numbers, especially when perceived as the only output. Therefore, applying them pervasively throughout an application can increase interaction complexity.

Non-use obtrusiveness for haptics comes down to the size and weight of the haptic interface elements, be they tactors or exoskeletons. To minimize tactor size and weight, provide power to each one from a common power source, negating the need for each tactor to have its own source of power.

Table 9-2 summarizes sources of OI with a haptics interface and the approaches discussed above to minimize or eliminate them.

Table 9-2. Transparent Haptics Interface Design Approaches

Operational Inertia Component	Sources of Operational Inertia	Design Approaches for Transparent Use
Setup Effort	• Placing a large number of tactors	• Embed them in a garment
	• Putting on an exoskeleton for force feedback	• Pre-attach skeleton to wearable substrate (ex. A glove)
Interaction Complexity	• Discerning activating tactors and semantics	• Ensure sufficient tactor separation to allow easy discernment of which is active
		• Place tactors on areas of the body that provides solid vibration base
		• Use a small number of

Operational Inertia Component	Sources of Operational Inertia	Design Approaches for Transparent Use
		vibration patterns for easy recognition and recall.
Non-use obtrusiveness	• Tactor size, weight	• Provide tactor power from centralized source

9.3 EYE TRACKING

Eye tracking measures the spatial direction (gaze and eye fixation) of where the eyes are pointing. It can help provide information of what the observer found interesting, and how the observer perceived the scene he was viewing. Eye tracking follows the path of an observer's visual attention [22].

Eye tracking systems can be classified as wearable or non-wearable, and further as infrared-based or appearance-based. Infrared-based systems employ a light shining on the subject whose gaze is to be tracked. The retina is a diffuse retro-reflector, so long-wavelength light, such as infrared, reflects off the retina and, upon exit, back-illuminates the pupil [26]. This produces a "red-eye effect" in the individual which the system uses to determine the direction of gaze.

Appearance-based systems use computer vision techniques to find the eyes in the image and then determine their orientation. These systems use both eyes to predict gaze direction, so the resolution of the image of each eye is often low, which makes them less accurate than infrared systems using one eye.

In a wearable eye tracking system, all of the equipment (cameras, illuminator, detector, etc.) are worn, typically on the head. Wearable (that is, head-mounted) eye trackers typically report the user's gaze point relative to the image obtained from a second camera which is fixed to the user's head, rather than the coordinates on a screen, as does a desktop or non-wearable eye tracker.

Eye tracking has not been used much in wearable systems since these are many issues that must be resolved before they are suitable. The main issues include intrusiveness, speed, robustness, and accuracy.

To understand the challenges that eye tracking faces, consider the requirements proposed in [28] for an ideal eye tracker:

- Offer an unobstructed field of view with good access to the face and head. This is crucial for a wearable system.
- Make no contact with the subject. This is the optimal for a wearable system, offering ideal non-use obtrusiveness.
- Artificially stabilize the retinal image to allow for optimal accuracy.

1. Accurately detect and track any movement of the eye. Depending on the level of analysis desired, the type of hardware and algorithms can vary widely. Gaze analysis can be performed at three different levels [29] including highly detailed low-level micro-events such as jitter and brief fixations, low level intentional events resulting in the smallest coherent units of movement that the user is aware of during visual activity, and coarse-level goal-based events.

- Provide a tracking range that spans the entire range of movement of the person's eye.
- Possess a real-time response. This is especially necessary for a wearable since the user may be in motion and his gaze may have to shift rapidly among objects in the environment.
- Measure all three degrees of angular rotation and be insensitive to ocular translation (smooth eye movement).
- Be easily extended to binocular recording. This may not be required for a wearable system.
- Be easy to use on a variety of subjects. Note that this is probably not necessary for a wearable system since most wearables are used solely by their owner.

Even allowing for those requirements that could be relaxed (accuracy, resolution, dynamic range, and response speed) and those that may not apply to a wearable system (use on a wide variety of subjects, binocular recording), the challenges facing eye trackers for a wearable system are formidable.

There are three technologies available to implement a wearable eye tracking system [27]. For a wearable system, the only viable technology is measuring the reflection of some light that is shone onto the eye. Typically, infrared light is used since it is invisible to the eye. The other technologies are much too invasive.

9.3.1 Devices

There are two broad categories of applications utilizing eye trackers [22]. Diagnostic applications utilize the eye tracker to provide information of the viewer's focus. The eye tracker is simply used to record eye behaviors for analysis, either real time or later offline. This group includes studies which test the effectiveness of some of some object on a screen, such as the location of a product in an advertisement movie. Of interest is how often and for how long the user's gaze fixates on the product under consideration.

Of more interest to wearable systems are interactive applications. One such application is 'Eye-aRe' [23]. Although not an eye tracker per se, the system contains an IR transmitter and receiver on a PCB board that attaches to one of the arms of a pair of glasses (see Figure 9-3). The transmitter contains an IR LED with a angle of transmission of about 20° that periodically transmits a unique code. This allows the system to determine when the user's head is oriented towards another user or towards a spot on a screen. This allows the system to detect eye contact with other Eye-aRe wearers in the environment.

Like other reflective systems, fluctuations in IR light are used to detect the user's pupil. An IR LED and a phototransistor at the front of the board are pointed inward towards the user's own eye. As the user's eye moves, the amount of IR reflected from the eye changes. A constant amount of reflection indicates that the user's eyes are fixed on an object in the environment.

Note that Eye-aRe assumes that the user's gaze direction is always in the direction the head is oriented. The system does not actually track the direction of the eye itself. Eye contact is determined when the user's head is oriented towards another user and the user's eye is fixated.

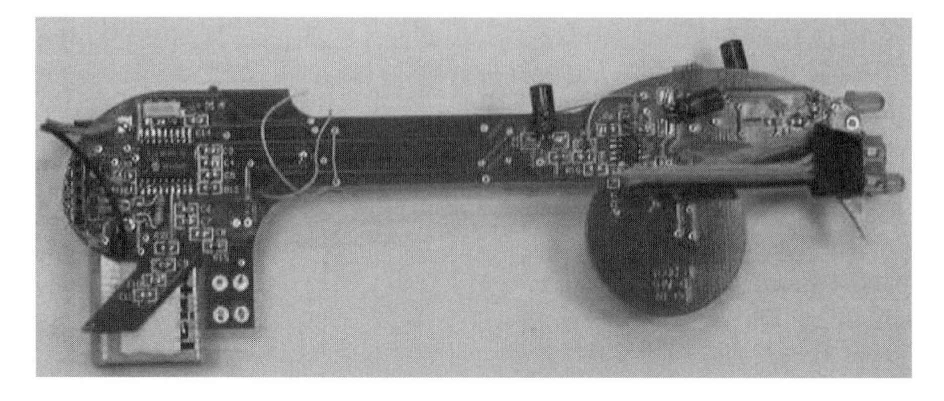

Fig. 9-3. The Eye-aRe Board *(Eye-aRe, Copywrite Ted Selker, Jorge Martinez, Andrea Lockerd Tomaz, Winslow Bureleson 1999)*

Much of the eye tracking is done on the wearable board itself. A major advantage to this approach is detection speed. A sample-and-hold circuit at 60 Hz is used to detect eye contact, and an onboard PIC micro-controller is used to detect fixations within the signal.

Another wearable eye tracking device is the ECSGlasses [27]. This system uses both a bright pupil and a dark pupil to track the user's point of gaze [30]. The system alternates flashing LEDs that are on the axis of the user's sight and those that are off axis. The flashing is at a rate coordinated with the camera's frame rate. As a result, the frames retrieved from the camera are alternating bright and dark. The camera is mounted between the wearer's eyes. This allows the system to detect eye contact with another person without requiring the other person to wear ECSGlasses, unlike Eye-R which requires both people to wear the augmented glasses.

In the final analysis, eye tracking does not currently seem to be a viable interface for transparent wearable systems. With the exception of Eye-R and ECSGlasses, the equipment worn is intrusive and bulky. Even Eye-R, the most wearable of the systems, is limited in that it assumes the user's eyes are pointed in the same direction as their head, which is clearly not always the case.

In addition, all of the eye tracking systems available today make the user look very geeky and none are aesthetically pleasing. However, eye tracking may prove to be a useful interface mechanism in the future.

9.4 MULTIMODAL USER INTERFACE

The main advantage of a wearable system over a desktop system is that it is always with the user and goes where the user goes. This means the wearable system is exposed to many different environments. Conditions in these environments include high ambient noise, hands busy, eyes busy, and multiple external stimuli competing for the user's attention. The system must adapt to these conditions and make it possible for the user to utilize the system regardless of the environmental conditions.

To do this, the wearable must support multiple UI mechanisms – speech, gesture, visual output, etc[94]. However, the ease of use of these interfaces is governed by:

- The ease of use of each interface mechanism separately
- The ease of switching among different interface mechanisms
- The ability to use multiple interface mechanisms concurrently in a synergistic manner.

It is the last element – concurrent, synergistic use of multiple interfaces that is the domain of a multimodal user interface.

To see how a multimodal user interface differs from standard interface mechanisms, let's look at the characteristics of the three types of interface interaction models.

[94] A note about terminology. We will refer to speech, gesture, haptics, etc. as UI mechanisms. A user interface will include one or more of these UI mechanisms.

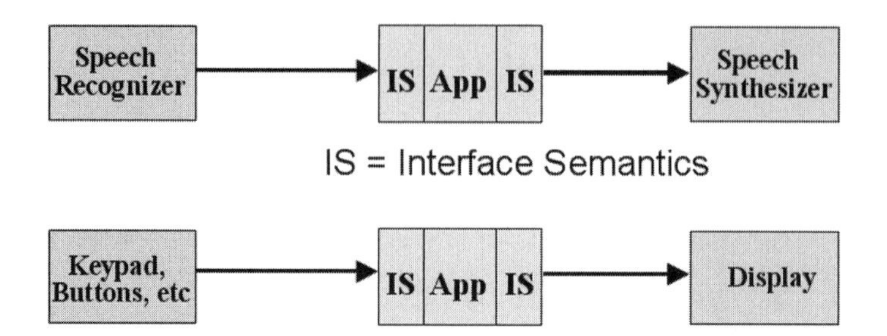

Fig. 9-4. Separate UI Interaction Model Architecture

9.4.1 Separate User Interfaces

In this interaction model we have multiple user interfaces supported by the wearable system. However, each UI is used separately and for a disjoint set of applications. There is no switching between interfaces within an application. This is the simplest, but least useful and effective interaction model for wearable systems. Figure 9-4 shows a simplified version of the interaction model architecture supporting speech and button/keyboard input.

The biggest problem is that as the environment changes, the current interface mechanism can become inappropriate and very difficult to use. Consider the user in the office utilizing a speech interface to dictate a letter. Under the mostly quiet conditions of the office the speech interface can be very effective. However, when the user moves outside, the high ambient noise due to traffic and general activity make the speech interface inaccurate and thus ineffective. Under the separate UI interaction model, the user cannot use any other interface with the application and looses the ability to use the application.

The problem is that the use of a specific user interface is hard coded into an application. Each application manages the input syntax and semantics of the specific UI it supports. This makes the interaction model very inflexible.

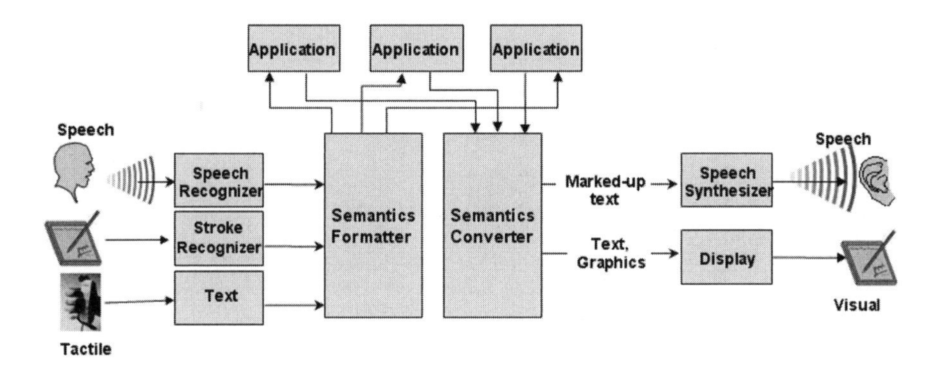

Fig. 9-5. Separate, Sequential UI Interaction Model Architecture

9.4.2 Separate Simultaneous Interfaces

Hard coding a specific user interface mechanism into an application is extremely limiting and is not good design practice. However, it makes the implementation of the interaction model quite simple.

We can relax the limitation at the cost of making the interaction model more complex. In the Separate Simultaneous Interfaces interaction model an application can use multiple user interface mechanisms, for example speech and GUI. However, there is no correlation or collaboration among the interfaces. This is perhaps the most common interaction model in current wearable systems.

In this model (Figure 9-5) the input and output dependencies are hidden from the applications. Each input is converted to a common device independent semantics format. Each application generates output using this common semantics format which is then converted into the proper device dependent format and sent to the output device.

While this model allows the user to switch from one interface mechanism to another while in the same application, it does not permit the use of multiple UIs concurrently in a collaborative fashion. The semantics have been separated from the applications into a common format, but there is no collaboration or combining of input from multiple UI mechanisms. Nevertheless, it is a big improvement over the Separate Simultaneous Interfaces interaction model.

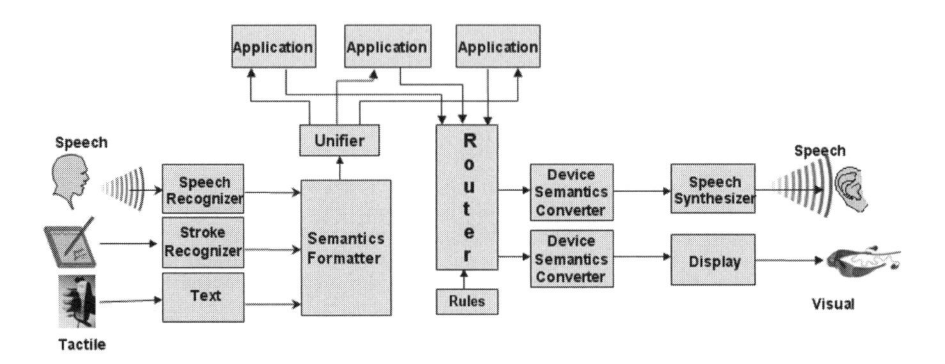

Fig. 9-6. Multiple Collaborating Interfaces Interaction Model Architecture

9.4.3 Multiple Collaborating Interfaces

In the multiple collaborating UI interaction model, the user can employ multiple UI mechanisms simultaneously in a collaborative manner to provide input to the wearable system. This has several advantages:

- The optimal interface can be used for the various elements of a command. For example, while speech may be the best input mechanism for most of a command, pointing on a tablet may be the most effective way to indicate a precise location as part of the command.
- The user can select among multiple elements that would otherwise be ambiguous. For example, if you must indicate a specific instance of a group that does not have an individual identity; you can give the majority of the command by speech and select the individual item via pointing.
- Using multiple UI mechanisms allows each mechanism to compensate for the limitations of the others [31]. The result is fewer errors.

The last point is especially significant. One of the major characteristics a wearable system must have is reliability. Manual error correction in particular can impose a large cognitive load on the user.

Figure 9-6 shows a simplified view of typical multimodal interface architecture employing the multiple collaborating UI interaction model. This architecture contains elements present in many multimodal UIs ([31], [33], [34]) and differs from the previous one in that the different simultaneous inputs are unified into a single input in the common semantic format before being passed to the applications. The applications are unaware of which or how many different user interface mechanisms were involved.

Unification is a defining feature of this interaction model of multimodal systems. Unification determines the mutual compatibility of input from two interface mechanisms, and if they are consistent combines them into a single result [35]. In [35] a multimodal integration agent determines and ranks potential unifications of separate spoken and gestural input.

The unifier agent fields incoming typed feature structures representing separate interpretations of speech and of gesture, identifies the best potential interpretation, multimodal or unimodal, and outputs a typed feature structure representing the preferred unification. The unification operation is performed over typed feature structures [36]. Input is considered complete when the type feature structure contains a full command specification and therefore does not need to be integrated with another input mode.

Every input is time stamped and time stamps from separate inputs must overlap within a specific time window to be considered part of the same command and be successfully unified. Completely filled in structures are then translated into application commands.

On output, information from an application is converted to the individual device dependent semantics format for the devices to which the information will be sent. The devices receiving the output are determined by rules. For example, one such rule could be to send output to all of the devices that are available and can render it [32]. If the output is text and both the speech synthesizer and display are available, the output would be sent to both.

These rules can incorporate context. For example, if both the speech recognizer and display are available, but the wearable system knows the user is in a location where speech output is inappropriate (say a movie theater), the text output would only be sent to the display.

As mentioned above, a multimodal interface can perform with fewer errors than the corresponding collection of separate interfaces because [31]:

- Users can select the input mode that they judge to be less error prone for particular lexical content. For example, they may choose to use text to enter a word the speech recognizer would have difficulty with such as a foreign name.
- The users' language is simplified when interacting multimodally, which reduces the complexity of natural language processing and avoids errors. Since they are using an additional input mechanism, speech input tends to be briefer, have fewer disfluencies, fewer reference pronouns, etc, all of which reduces the chance for error from the speech recognizer.
- Users tend to switch modes after system errors, often choosing to employ more suitable interface mechanism which facilitates error recovery,
- Architectural constraints imposed by semantic unification in multimodal systems often rule out incompatible combinations of input choices returned by the interface mechanisms (for example speech and gesture). This allows the system to choose the correct input even though it is ranked lower in the N-best list from the input mechanism. For example, sweeping out an area with a pen while giving the spoken command to show all friends within that area provides additional information that can help to pick the correct utterance from the list of N best utterances returned by the speech recognizer, even if that utterance is not the one judged most likely by the speech recognizer.

9.4.4 Toward Transparent Multimodal User Interfaces

Designing multimodal user interfaces is still much of an art. Nevertheless, Table 9-4 lists some of the sources of OI that must be addressed if we are to design transparent multimodal interactions. Much of this is derived from [37].

Table 9-4. Design Approaches for Transparent Multimodal Interaction

Operational Inertia Component	Sources of Operational Inertia	Design Approaches for Transparent Use
Setup Effort	• Configuring interface mechanisms and their	• Minimize the setup effort of each UI mechanism in

Operational Inertia Component	Sources of Operational Inertia	Design Approaches for Transparent Use
	interaction	isolation • Use rules and context information to configure the interaction among multiple UI mechanisms
Interaction Complexity	• Inappropriate UI mechanisms used for the current cognitive load experienced by the user	• Allow the user to select the user interface mechanism (gesture, text, etc) based on the user's current situation to ensure the cognitive load is within the current CLB
	• Non- synergistic splitting of output among multiple mechanisms	• Information presented in each mechanism should be mutually reinforcing • Use context information to ensure multiple output mechanisms render information synergistically
	• Excessive delay among coordinated output to different mechanisms	• Ensure coordinated output is provided within a small temporal window of one another
	• Inconsistent state of user interaction among different mechanisms	• Provide a global state manager to ensure the same user state is seen among all interface mechanisms
	• Differing command sets among devices using same interface mechanism	• Provide centralized command policy manager for each mechanism
	• Confusion about what interface mechanisms are available and/or appropriate in the current context	• Allow user to choose which UI mechanisms to use for input. System selects most appropriate one to use for unsolicited output.
Non-use obtrusiveness	• Rendering output to an interface mechanism that is inappropriate or ineffective for the current	• Employ rules based on current context or user preferences to autonomously or semi-

Operational Inertia Component	Sources of Operational Inertia	Design Approaches for Transparent Use
	context	autonomously determine which mechanisms are used for input and output

We discuss each of these sources of OI and possible design approaches for their minimization below.

- **Configuring interface mechanisms and their interaction:** refer to past sections in this and the previous chapter on the individual UI mechanisms (speech, gesture, GUI, etc) to minimize each mechanism's setup effort in isolation.

  Also, employ user preferences, context information, and any explicit user directives to configure required interactions among multiple UI mechanisms. This is done at system startup, but may also have to be refined dynamically with changes in context.

- **Inappropriate UI mechanisms used for the current cognitive load experienced by the user:** allow the user to select the user interface mechanism (gesture, text, etc.) that imposes the least cognitive load for the user's current environment and primary task. The system can monitor the user's performance and make available (but do not require) the use of alternate, more effective interface mechanisms given the current context.
- **Non-synergistic splitting of output among multiple mechanisms:** minimize the number of mechanisms to which the user must simultaneously attend in order to comprehend the material being presented. In cases where multiple mechanisms (say speech and visual display) are used, the information presented in each mechanism should be mutually reinforcing

  Provide UI affordances that maximize the advantages of each mechanism to reduce user's memory load. For example, spatial information is usually best entered using a spatial analog such as positioning pen on a location on a tablet or sweeping out an area. In contrast, audio, including speech, is often better in presenting state information, serial processing, attention alerting, or issuing commands.

Using context information can help ensure that the multiple output mechanisms being used render the information synergistically as suggested below.

- **Excessive delay among coordinated output to different mechanisms:** Ensure system output mechanisms are well synchronized temporally. Provide coordinated output to each interface mechanism within a small temporal window of one another. For example, if presenting information both visually and by speech and each mutually reinforces the other (as recommended above), make sure they are presented together. Otherwise much of the mutual reinforcement is lost and user's cognitive load, and with it the interaction complexity, is increased.
- **Inconsistent state of user interaction among different mechanisms:** Ensure the current system interaction state is shared across interface mechanisms. Provide a global state manager to ensure the same user state is seen among all mechanisms. This will make it easier for the user to switch among mechanisms as necessary or desired. It will also reduce POLA[95] violations and make the interface more predictable, keeping interaction complexity low and increasing the user's trust in the system. Display information to support users in choosing alternative interface mechanisms as required.
- **Differing command sets among devices using same interface mechanism:** Interface mechanisms (e.g., speech, gesture, etc.) should share common features, such as presentation and prompt terminology as much as possible and should consistently refer to tasks using the same terminology across mechanisms. This promotes the user's view of the multimodal interface as a single, integrated, highly functional user interface. At a minimum, design commands that are consistent and

[95] Principle Of Least Astonishment. See 5.1.2.

mutually reinforcing across all input devices of the same mechanism. Provide centralized command policy manager for each mechanism (see Chapter 5).

- **Confusion about what interface mechanisms are available and/or appropriate in the current context:** Ensure users know which mechanisms are available to them. They should be aware of alternative interaction options without being overloaded by lengthy instructions that distract from the primary task. Even better, allow user to select any input mechanism desired. Have the system select the most appropriate output mechanism for unsolicited output.
- **Rendering output to an interface mechanism that is inappropriate or ineffective for the current context:** multimodal interfaces should adapt to different contexts of use. Employ rules based on current context and user preferences to autonomously or semi-autonomously determine which interface mechanisms should be used for output. This includes conditioning the quantity and method of information presentation on the current context, the user's facility and preferences with the interface mechanisms, and device capabilities.

9.5 INTERFACE PERSONALIZATION

Karla looked at the night sky, its blackness perforated with millions of stars. Its immenseness overwhelmed her and she shuddered. However, she felt comforted by the thought – the certain knowledge - that in all the vast expanse of the universe, she was unique. Special, one of a kind. Not important or famous perhaps, but unique nonetheless.

But, the universe is cold, uncaring. And it cared nothing for Karla's belief – indeed her need – that she was unique. And so, Karla, whose full designation was Karla 5-M3, was to learn the awful truth. A truth that would forever change for her who she was and what she would become. ...

Individuality. It is perhaps the characteristic we count on the most to give us a sense of worth. Throughout history, men and women have adorned themselves in an effort to highlight their individuality. Even when conformity is highly prized, people find subtle and creative ways to display their uniqueness. Identical twins – cut from the same genetic cloth –soon begin to display differentiating traits and behaviors. Though they are very much alike, their individuality is never questioned. This sense that we are unique is so essential to our concept of being human that we can scarcely conceive of its not being true.

However, expressions of individuality are limited in current computing and communications devices. Pictures of family and images expressing personal interests appear on desktop PCs and laptops. Cell phones are personalized with ring tones and antenna lights and colorful covers. Newer large screen phones are starting to sport wallpaper images and display graphical themes like their PC counterparts.

But this is a mere fraction of what will be possible with wearable systems and pervasive computing environments. Indeed, the very notion of an interface, its characteristics, and even the concept of personalization will change with these new devices and environments.

As we have discussed in Chapter 6, computing is ubiquitous in a pervasive computing environment. In these environments computing elements are in every type of device – from a light switch to a door jamb. The computing is invisible and non-intrusive. You will often be unaware of the computing except for the more effective operation of the device.

Wearables will also conduct much of their operation non-intrusively with little user awareness. Wearable devices and applications are designed not to require the user's full attention when being used.

The hidden nature of computing in pervasive environments and the emphasis on non-intrusive operation of wearable applications would at first glance seem to limit the opportunities for user interfaces and their personalization. After all, there is not much of an interface for a lamp or a body sensor. These devices are *microClients* – devices with limited functionality and computational complexity. Many devices in a pervasive computing environment will be microClients. The potential for significant interface personalization with most microClients is very limited but it is there.

Nevertheless, the potential for interface personalization with wearables is great. This will be driven by three characteristics of wearable systems:

- The intimate nature of the systems interactions with its user;
- The inherent multi modal interfaces required for effective system operation;
- Ad hoc communication with devices within the user's Personal Operating Space;

Wearable systems are an intimate technology. That is, in their mature form they will exist in a close symbiosis with the user. They will monitor the user's activities, body state, and characteristics of the environment in which the user is immersed. This data will be used to infer state and synthesize information that will allow the wearable to be proactive and highly targeted in the assistance and information it gives the user. The user will come to regard the wearable as a personal extension. The user will want to extend his sense of identity to the wearable.

Wearables will typically be with the user and operating the entire time the user is awake. They will be used in all of the different situations and environments the user experiences. Because of this, effective operation of the wearable requires multiple user interfaces. The multiplicity of interfaces and their potential simultaneity create new opportunities for personalization.

As the user moves around in the performance of their daily activities, they will encounter client devices in an unplanned, ad hoc manner. The ability to immediately and easily use these devices is a hallmark of pervasive computing and transparent wearable systems. However, this poses problems of using these devices effectively since each may have its own user interface. Currently, it would be up to the user's wearable to ensure that its information is properly formatted to meet the specifications and constraints of the interface on the client device. By reversing this relationship, we not only ease the burden of the wearable device to use the clients in an ad hoc manner, we also open up many new opportunities for interface personalization.

What will the concept of a user interface be in such an environment? The wearable system user interface will bear little resemblance to that of a PC. Instead, it will be associated with the person. A user interface will be the person's window (note the small "w") into the world full of devices that do computation (however little) and communicate with the user. This has several interesting implications:

- The notion of a "standard" user interface will be much less universal and much more personal. The number and diverse types of devices with which we will interact makes a detailed universal interface standard for mainstream wearables system unwieldy and impractical. Instead, each person will define his or her own interface (or interfaces) through which they communicate with the environment around them. The interface will be "standard" only to them. This will be especially true for very simple devices, which will contain little or no interface libraries, but will instead 'inherit' the interface from the user's wearable system for the duration of the interaction as their capabilities allow.

  Instead of fully specified standards like Windows, we will have high-level guidelines that capture well-established design principles dealing with effective spatial relationships, the psychology of color, and human visual and auditory perception. These guidelines will also include suggested ways of combining multiple interface mechanisms such as graphics and speech together. These guidelines will be incorporated into a UI expert system that will apply the extensive user personalization preferences in a cognitively effective and aesthetically pleasing presentation.

- The intelligence for the interface will reside in the person's wearable system's central unit, not in the device with which the user is interacting. This will even be true when interacting with a PC. The PC will no longer impose its user interface upon the wearable. Rather, it will receive the user's interface specifications and use basic presentation routines to display the information. This will make the user – PC collaboration much more effective.

  This is where the reversal of the current relationship will be most apparent. The current interfaces like Windows and the Mac OS interface are driven by the fact that the PC is not a personal device. PCs are devices that are often shared among multiple people, or deployed in large numbers to members of an organization who do not have ultimate control of their computing resources. This means there is a real benefit to having a detailed standard that limits the amount of alteration possible. This facilitates sharing PCs among members of the organization and makes maintaining and servicing large numbers of them easier.

Contrast this with a wearable that is intimately personal. The user owns the system and it is directed at helping only the user. Therefore, it is more important that the wearable system satisfies the user's wishes than that it promote the sharing of the system with others. As a result, there is less benefit to be gained by a rigid, detailed interface standard[96].

- On the other hand, many devices that are not worn on the body will be shared *in a personal manner* by multiple people. In a mature pervasive computing environment, the cost of many microClients will be so low that there will be little incentive to enforce ownership. Today when we buy a pack of paper we do not jealously guard ownership of each sheet. Indeed, it is not uncommon to leave pieces of paper out where others can use them. The low cost of the paper makes it impractical to ensure that no one else uses it. Now assume electronic paper reaches the same low cost. We would leave sheets of it lying around after we are done using it. Someone else would encounter it and use it. This opportunistic use of microClients will be common in a mature pervasive computing environment.

  This means the same device would become a part of many people's wearable system. However, each person would have exclusive, personal use of the client for a period of time. The person's wearable system will send the user's interface specification to the client[97]. The client will apply

[96] The one argument against this, of course, is that the lack of a standard interface will make getting help with the operation of the wearable system from a remote support person harder. Customer support people will be unable to simply read from a script like they do now to instruct users on how to operate their wearable. This is another reason why it is crucial that wearable systems have very little setup effort and interaction complexity. Wearable systems must be Near Zero Operational Inertia Devices (NZOIDs). NZOIDs are extremely easy to configure and set up and they are almost transparent to use.

[97] Many microClients will simply provide a bitmap. In this case the wearable system's central unit will acquire the physical display characteristics from the device and construct the

the user's interface specifications to the limit of its capabilities. Since the device is, temporarily, part of the user's wearable system and it is being used in an exclusive, personal manner, there is significant benefit to implementing the user's personalized UI rather than enforcing adherence to a 'standard" interface specification.

- The notion of a user interface will expand to include many interface mechanisms, including speech (input and output), graphics, pen input, gesture, and even biometrics and smell. Wearable devices will support multiple mechanisms in a seamless overall interface. The choice of which mechanisms to use among the set and how they should be combined will be driven by the user's own UI specification and from the UI expert system employing the guidelines discussed earlier for multi modal UIs.
- The notion of preferences will expand beyond its current set of simple GUI characteristics (window color, icon font, background image, sounds, etc.). A person's UI preferences will cover all of the available interface mechanisms and topics such as how help is provided and errors handled. As an example, we discuss two emerging interfaces: speech and gesture. Similar personalization opportunities will exist for graphical, pen, and other interfaces.
  - Speech
    - *Output voice:* Pure synthesis Text To Speech sounds unnatural, although it is quite intelligible. Newer TTS systems use concatenated synthesis in which human recorded speech segments are concatenated together. This gives completely natural human quality speech. You will be able to get a TTS library of your own voice or even purchase libraries of other voices, much as you can purchase ring tones for cell phones today.

---

display itself. The wearable system will then send the bit stream to the microClient, which simply renders the bitmap. High speed PANs will support the data transfer speeds required.

- *Speaking parameters* (pitch, speed, etc): These elements of speech are called prosody. Prosody adds inflection and conveys much meaning in verbal communication. The same sentence, said with different inflection, can mean totally different things. You will be able to apply "emotion templates". An emotion template will specify the prosodic elements that will be applied to your system's verbal output to portray a specific emotion.

  Using the awareness of your current emotional state, an emotion template will be chosen to either be consistent with or attempt to alter your current emotional state. For example, if you are upset and currently driving, your wearable system may adopt an especially soothing emotional template when it verbally interacts with you (if your preferences have indicated that you want it to do this).

- *Verbal interaction style* (i.e., how casual the interaction with the device should be): Verbal interaction style is a higher level of personalization than voice type or prosody. It reflects the use of language including idioms. The choice of words and phrases convey the level of formalism. There may be cases in which you want the wearable to be formal and precise, for example, when providing error messages or messages of a serious nature. Other times a more casual style will be preferred since it may be easier to listen to.

The three previous elements (voice, prosody, and interaction style) are part of a *personality*. Such personalities will become common and may even be offered for sale. Use of these personalities will further reinforce the concept of an identity for the wearable and promote the relationship between it and the user. We discuss personalities in more detail in Chapter 11.

- *Mechanism for acquiring the user's attention prior to rendering verbal output:* Before the wearable system can provide information to the user, it must ensure the user is paying some attention to it. While we don't want to focus our entire attention on the wearable, we must be aware that it is addressing us. Since the wearable will be operated hands free in many cases, it will be competing with all of the other audio and visual stimuli to which the user is exposed[98].

  Mechanisms to acquire the user's attention include preceding messages with the user's name, emitting a distinctive tone or phrase, and vibration. Each of these are appropriate in some situations and inappropriate or less effective in others. The wearable system, using context information and the user's preferences and notification specifications, will select the most appropriate mechanism to maximize the probability of getting the user's attention while minimizing the intrusiveness of the notification. The user will be able to specify the set of notification mechanisms and even provide some guidance as to when they should be used.

- Gesture
  - *Gesture Thresholds:* Most of us use our hands to help convey what we are saying. Each person has a different level of expressiveness when speaking. Much of this is done unconsciously or nearly so. A wearable system gesture interface must take this into account. It must distinguish when the user is giving a command via a gesture and when the gesture is just the normal hand motions that accompany speaking. In many cases the user will use both a speech interface and gesture interface simultaneously. This means that sometimes the same gesture may be an unconscious motion during speech and at other times it will be a specific gesture interface command given while the user is speaking.

[98] Recall the discussion in Chapter 3 on modes of user – system interaction.

To differentiate between the two cases, the user will specify gesture thresholds. These specify the minimum amplitude of a gesture that is to be taken by the wearable as a gesture interface command. If the gesture is done with less than the specified amplitude, it is regarded as an unconscious movement of the hands[99].

- *Gesture contours:* Everybody makes the same gesture a little differently. The gesture contour is based on the person's size, gender, how much they use their hands when talking, etc. The user will train the gesture interface to recognize their specific variations of the gesture they want to associate with a command.

  In some cases, the contour and/or amplitude of the gesture will be conditioned on the user's context. For example, the same gesture made in a car will be different when I am standing up. The knowledge of the user's context will be used to apply the correct gesture thresholds and contours.

The information contained in this expanded user interface is extensive and dynamic. New elements will be added and current elements modified at any time. To prevent time consuming and error prone manual configuration, the wearable system must be aware of the user's environment, actions, and preferences. The system will learn what information is required for the interface. With the exception of an initial manual configuration, this learning capability will greatly reduce the user-initiated maintenance and set up effort of the interface.

We have focused on the interaction of the wearable system with the user. There is another, equally important aspect of the use of wearable systems:

[99] Recall for the discussion in this chapter on gesture interfaces that the interface could be monitoring the user's gestures while the user is speaking for unconscious gestures correlated with speech intonation to help recognize what the user is saying. This makes it all the more important that conscious gesture commands be distinct or used in much different contexts from unconscious, speech correlated gestures.

the interaction with society at large. Social issues surrounding the wearable system can be very difficult to resolve. This is the focus of the next chapter.

REFERENCES

[1] What's a Theremin?, 2005, Theremin World, http://www.thereminworld.com/article.asp?id=17

[2] Phiffer D. and Zer-Aviv M., 2006, Atlas Gloves, A DIY Gesture Interface for Google Earth, http://atlasgloves.org/

[3] Nielsen M., Störring M., Moeslund T. B., et. al., 2003, A Procedure For Developing Intuitive And Ergonomic Gesture Interfaces For Man-Machine Interaction, Technical Report CVMT 03-01, Aalborg University

[4] Cassell J.,1998, A Framework For Gesture Generation And Interpretation, Computer Vision in Human-Machine Interaction, Cambridge University Press

[5] McNeill, D., 1992, Hand and mind: What gestures reveal about thought. Chicago: University of Chicago Press.

[6] Statner T, Auxier J., Ashbrook D., et al, 2000, The Gesture Pendant: A Self-illuminating, Wearable, Infrared Computer Vision System for Home Automation Control and Medical Monitoring, Fourth International Symposium on Wearable Computers, pp 87-94

[7] Sharma R., Yeasin M, Krahnstoever N, et. al., 2003, Speech–Gesture Driven Multimodal Interfaces for Crisis Management, Proceedings Of The IEEE, Vol. 91, No. 9

[8] Boff, K.R., Kaufman, L., and Thomas, J.P. (Eds.)., 1986, Handbook of Perception and Human Performance: Sensory Processes and Perception. Vols. 1 and 2. Wiley, New York, N.Y

[9] Tan H. Z. and Pentland A., 1997, Tactual Displaysfor Wearable Computing, Personal Technologies, Springer – Verlag, Vol 1, pp. 225 – 230.

[10] Burdea G. C., 1999, Haptic Feedback for Virtual Reality, Proceedings of International Workshop on Virtual prototyping, Laval, France, pp. 87-96, May

[11] Immersion Corporation, 2003, CyberGrasp™ v1.2 User's Guide, http://www.immersion.com/3d/docs/CyberGrasp_030619.pdf

[12] Dennerlein J., Becker T., Johnson P., et. al., 2003, "Frustrating Computer Users Increases Exposure to Physical Factors," Proceedings of the International Ergonomics Association, Seoul, Korea, August 24-29

[13] Pompei F. J., Sharon T., Buckley S. J., et. al., 2002, An Automobile-Integrated System for Assessing and Reacting to Driver Cognitive Load, 2002-21-0061, Society of Automotive Engineers, Inc.

[14] MacLean K., design of haptic interfaces, 2001, UIST 2001, http://www.cs.ubc.ca/~maclean/

[15] Gemperle F., Ota N., and Siewiorek D., 2001, Design of a Wearable Tactile Display, Proceedings of the 5th International Symposium on Wearable Computers, Zurich, Switzerland, 7--9 October 2001

[16] Bloomfield A. and Badler N. I., 2003, A Low Cost Tactor Suit for Vibrotactile Feedback, Computer & Information Science / Technical Report 2003, University of Pennsylvania

[17] Lindemanl R. W. Page R., Yanagida Y., et. al., Towards full-body haptic feedback: the design and deployment of a spatialized vibrotactile feedback system, Proceedings of the ACM symposium on Virtual reality software and technology (VRST), pp. 146 - 149

[18] Logitech IFeel Optical Mouse, 2001, http://www.hardware-one.com/reviews.asp?aid=221&page=5

[19] Cardin S., Thalmann D., and Vexo F., 2005, Wearable Obstacle Detection System for visually impaired People, HAPTEX 2005, Hanover, Germany, December 1, 2005

[20] Ito K., Okamoto M., Akita J., et. al., 2005, CyARM: an Alternative Aid Device for Blind Persons, Proceedings of Computer-Human Interaction (CHI 2005), pp. 1483-1486

[21] Sound Foresight Ltd, 2006, UltraCane, http://www.soundforesight.co.uk/index.html

[22] Djeraba C., 2005, State of the Art of Eye Tracking, Technical report LIFL – 7-2005, Université des Sciences et Technologies de Lille

[23] Selker T., Lockerd A., and Martinez J., 2001, Eye-R, a Glasses-Mounted Eye Motion Detection Interface, CHI'01 Interactive Posters

[24] Duchowski, A. T., 2002, ``A Breadth-First Survey of Eye Tracking Applications'', Behavior Research Methods, Instruments, & Computers (BRMIC), 34(4), November 2002, pp.455-470.

[25] Smith J., D., 2005, ViewPointer: Lightweight Calibration-Free Eye Tracking for Ubiquitous Handsfree Deixis, Queen's University Kingston, Ontario, Canada

[26] Babcock J. S., Pelz J. B., and Peak J., The Wearable Eyetracker: A Tool for the Study of High-level Visual Tasks, Proceedings of the Military Sensing Symposia Specialty Group on Camouflage, Concealment, and Deception, Tucson, Arizona

[27] Glenstrup A. and Angell-Nielsen T., 1995, "Eye Controlled Media, Present and Future State," Technical Report, University of Copenhagen

[28] Hallett, P. E. (1986), Eye movements, in K. Boff, L. Kaufman & J. Thomas, eds, `Handbook of Perception and Human Performance I', pp. 10.25-10.28.

[29] Campbell C.S. and Maglio P.P., 2001, "A robust algorithm for reading detection," ACM Workshop on Perceptive User Interfaces

[30] Tomono A., Iida M., and Ohmura K, 1991, Method of detecting eye fixation using image processing. U.S. Patent: 5,818,954, 1991. ATR Communication Systems Research Laboratories.

[31] Oviatt, S. L.,. "Mutual disambiguation of recognition errors in a multimodal architecture", Proceedings of the Conference on Human Factors in Computing Systems (CHI'99), ACM Press: New York, 576-583. 1999

[32] Marsic I. and Dorohonceanu B., 2003, Flexible User Interfaces for Group Collaboration, International Journal Of Human–Computer Interaction, 15(3), 337–360

[33] Cohen P.R., Johnston M., McGee D.R., et al, 1997, QuickSet: Multimodal Interaction for Distributed Applications," Intl. Multimedia Conference, `97, 31-40

[34] Robbins C. A., 2005, Extensible MultiModal Environment Toolkit (EMMET): A Toolkit for Prototyping and Remotely Testing Speech and Gesture Based Multimodal Interfaces, PhD. Thesis, Department of Computer Science New York University

[35] Johnston M., Cohen P. R., McGee D, et al, 1997, Unification-based Multimodal Integration, ,Proceedings of the eighth conference on European chapter of the Association for Computational Linguistics, 281 – 288

[36] Carpenter, R. 1992. The logic of typed feature structures. Cambridge University Press, Cambridge, England.

[37] Reeves L. M., Lai J., Larson J. A. Et al, 2004, Guidelines For Multimodal User Interface Design, Communications Of The ACM January 2004/Vol. 47, No. 1, pp. 57-59

PART 5: SOME TOUGH HURDLES AND THE FUTURE

Chapter 10

SOCIAL ISSUES OF WEARABLES

Wearable systems promise to help us perform our tasks more efficiently and more enjoyably. Advances in technology will make wearable devices smarter, smaller, and easier to use. At the same time, the use of these systems will raise several social and cultural issues. Some of the potential social benefits of wearables have already been discussed. For example, Steve Mann has discussed their potential for empowering the user in terms of privacy and access to information [1]. This chapter is concerned with issues involving social conventions, personal feelings, and expectations involved in the use of mainstream wearable systems. Resolving these issues will allow people to utilize these systems to their full potential. Not resolving them could significantly impede and delay their widespread adoption.

10.1 WEARABLES AND THE ACCOMMODATION OF SOCIAL VALUES AND CONVENTIONS

Social conventions embody a common consensus of how we are supposed to act in public (and in some cases, private) situations. They define accepted rules of social interaction [2].

The use of conventions simplifies our lives. We don't have to reestablish rules of conduct every time we engage in specific activities. Failure to follow these conventions disrupts the normal flow of social interaction. We can appear rude, antisocial, and even dangerous.

The set of conventions in a culture is not static. It changes as the society changes, in an attempt to codify new patterns of social interaction. In many cases technology can define new patterns of interaction or behavior that eventually become established as new or modified conventions. We can expect social conventions governing human to human communication to be affected by the adoption of wearable systems among the general population, much as they did when the telephone became widely used [3].

In this section we consider how the use of mainstream wearables can conflict with various social conventions or values.

10.1.1 Conversing With Oneself

The near zero setup effort and unobtrusiveness of an NZOID as part of a mainstream wearable system means that there will be few visual cues that its owner is using it. This can be an issue when the user is in a public location such as an airport and is using the speech interface, for example, in taking a call. In these situations, the user will be perceived as violating a strong cultural convention: that of not talking to oneself.

This taboo may be breaking down in an example of society adapting to a pervasively deployed technology. More and more people are using wired headsets and, more recently, wireless headsets based on Bluetooth to speak on their cell phones hands free. This is making people more familiar and comfortable with the sight of a person speaking to no one around him.

However, even in these cases there are visual cues that can confirm the viewer's guess as to what the speaker is doing. In most cases, the user is holding the cell phone, just not up to the ear. The presence of a wire from the phone to the ear is a strong cue that the person is on the phone. Even in the case of a wireless Bluetooth headset, the headset is usually visible, even looking at the speaker head on. This is also a very discriminating cue that the user is on the phone.

But now suppose we remove all of those cues. Suppose the phone is in the user's pocket and not in his hands. Further suppose the Bluetooth earpiece is not visible outside the ear[100]. Now how is the speaker perceived by those around him? They will look at him, expecting to see one of the cues indicating that he is on the phone. Seeing none, they may be quite confused since the expectation now has become that such a person is on the phone. Seeing none of these cues, they may revert back to the response people had to this situation before cell phones and headsets became common. They may assume the user is indeed speaking to himself.

Now let's add the use of a gesture interface. The use of clearly visible gestures will further call attention to the person. While observers may not think the person is talking to himself, some may find this behavior inappropriate in public.

One way of addressing this is to define standardized gesture vocabularies. Then, when you see a person using one of these gestures you can assume they are interacting with their wearable system. However, the requirement to use specific gestures for specific purposes goes against the interface personalization we discussed in the last chapter.

10.1.2 Personal Privacy

Today one of the biggest social issues with cell phones is privacy, both the user's and those around him. From the recent experiences with camera cell

[100] This is not as outlandish as it seems. The Motorola Mini Blue Bluetooth earpiece [16] barley extends out of the ear. Most of it is contained just outside the ear canal in the ear. As time goes on, these devices will get even smaller.

phones, it is clear that our laws are behind our technology and that wearable systems will test the limits of this issue.

Transparent wearable systems have characteristics that can exacerbate the privacy issue. First, they are likely to have significant disk storage capability. Disk are on their way toward terabyte capacity [4], and with that capacity the wearable system will be capable of storing video of every waking moment of your life for several years on a single 3.5 inch drive. This data can be archived indefinitely. It can also be digitally manipulated and altered to create a completely different context and impression.

Second, the recording process is likely to be mostly invisible to those around us. Cameras are very small and can be operated using contextual clues with little intervention by the user or simply left running. The camera can be separate from the main wearable unit and can be placed where it can record without us having to handle it.

Third, they will go everywhere we go. Since they are worn, they will be considered part of us and thus may be afforded a certain amount of deference. This issue is more fully discussed in the section on reverence for personal space below.

In addition, wearables can use this data to, in real time and in the current context, retrieve information and use it for the user's benefit, perhaps at the expense of the person with whom the user is interacting. This particular point is discussed more thoroughly in the section on transparency and attention in face to face interpersonal interactions below.

Three scenarios in [5] illustrate some of the situations in which privacy concerns with wearables can arise. Among them are:

1. Using the wearable and vision prosthesis and processing the information to assist the user in navigating and performing his tasks. Since it is a prosthesis, it must go everywhere the user goes and operate all the time.
2. Visiting a person in their house, perhaps for the first time. The home is a place afforded special privacy protection and coming into one's home with a camera that is recording everything can cause privacy concerns.
3. Recording people in public and then obtaining information on them from the internet and other information sources. This information can be used

and manipulated for various good or bad purposes without the subject being aware.

4. Confiscation of the wearable system and search of the information gathered. This issue is discussed more fully in the section on reverence for personal space below.

Once wearable systems become essential cognitive and sensory aids, people with physical impairments may be allowed to use them legally in places where people without physical impairments can not. [5]. There is precedence for this in other areas. For example, seeing eye dogs are allowed in areas where other dogs are not such as restaurants, public buildings, and housing [6]. In the same way, laws may be passed or amended (for example, the Americans with Disabilities Act (ADA)) to allow user's to wear their systems in places that abled people may not be allowed to[101].

Arriving at a person's home at their invitation for a social gathering and wearing your wearable system with its recording and information processing systems active may become unacceptable. There is a large body of case law that has established the special level of privacy for the home.[102] For example, many states now criminalize home voyeurism, where a person views activity in another home without being invited in. While most of these statutes require physical trespassing (entry of the grounds without permission), some states have criminalized viewing that does not involve trespassing [7]. Although most of these laws were passed to address the issue of "peeping toms" viewing people in various stages of undress, they could be revised to address the detailed recording of a person's home, even if you are an invited guest. Just imagine your reaction if one of your guests

[101] While the ADA does not specifically mention service animals, the Department of Justice regulations that implement the ADA do [6].

[102] The opinions and conclusions expressed herein are solely those of the author and not those of any entity with whom the author is or has been affiliated. The author is not a lawyer and has had no legal training, and the conclusions and opinions regarding legal issues should not be taken as legal advice.

brought and operated a video camera the whole time they were at your house without your invitation or consent.

There is much less expectation of privacy in public places. Indeed, courts have held that persons in a public space voluntarily waive their right of privacy for their actions while in public [8]. However, if the recorded data was manipulated to portray a false impression of the person or their actions, that would be considered an invasion of privacy [9].

Note, that as long as the event is recorded and rendered accurately, even highly embarrassing events that occur in a public space are not protected, if those events are illegal. For example, in 1996 Brian Bates, an Oklahoma City resident , began video taping prostitutes and their customers having sex in public places in his neighborhood, then dialing 911 and placing the couples under citizen's arrest until officers arrived [10]. While several of the customers filed civil lawsuits for invasion of privacy, no criminal action was brought against Bates.

The issues of privacy with wearable systems will only get more involved as these systems are adopted and used in ways we have yet to conceive.

10.1.3 Reverence for Personal Space

An issue closely related to privacy is reverence for personal space. As NZOID wearables become integrated into people's clothing and jewelry and attached to the body, society may regard them as part of the person. This could convey upon them a legal status that current, non-NZOID systems do not enjoy.

The precedent for this may lie in a Supreme Court decision, Wyoming v. Houghton, decided on April 5, 1999 [11]. The court decided that a police officer, having probable cause, can make a warrantless search of a passenger's belongings (property), even if the evidence of a crime is associated only with the vehicle's driver.

More to the point, the court held that items that are carried, such as a purse, can be searched but a search of the person would be a violation of the fourteenth amendment guaranteeing freedom from unwarranted search:

> "The degree of intrusiveness of a package search upon personal privacy and personal dignity is substantially less than the degree of intrusiveness of the body searches at issue in United States v. Di Re, 332 U.S. 581, and Ybarra v. Illinois, 444 U.S. 85." [11]

And the court, in the referenced case, United States v. Di Re, 332 U.S. 581, stated:

"We are not convinced that a person, by mere presence in a suspected car, loses immunities from search of his person to which he would otherwise be entitled." [12].

Further, in [12] the arrested person had ration coupons stuck between his shirt and underwear. The court considered the search for those coupons to be a search of his person and thus protected by the Fourth Amendment to the U.S. Constitution

The court considered the articles between a person's clothes and the body part of the person and so may not be searched without a warrant. Thus, NZOID based wearables integrated into clothing or on the body may be off limits unless a warrant allowing a personal search was obtained[103].

This may support Mann's assertion that wearables will empower their users with protection from surveillance [1]. However, they will do this best if they can be considered part of the person.

[103] Again, the opinions and conclusions expressed herein are solely those of the author and not those of any entity with whom the author is or has been affiliated. The author is not a lawyer and has had no legal training, and the conclusions and opinions regarding legal issues should not be taken as legal advice.

10.1.4 Social Politeness

We all live in many social environments – work, school, gathering with friends, even just walking down the street. Many of the social conventions we are expected to observe are to manage and foster these social interactions for the benefit of all concerned.

A wearable system can certainly enhance our social connectedness. For example, a wearable system can provide easy, anytime access to social networking applications and web sites such as Dodgeball.com which, given your location will tell you who and what is around you [13].

However, as we have learned from our experiences with cell phones, wearable systems can also present challenges to the continued observance of these social conventions.

If our wearable systems do indeed support transparent use, the frequency and amount of communication with those remote from us will increase. One source of this increased communication will be the ability to engage in Opportunistic Communication.[104]

For a wearable system to manage a user's communication transparently, it must resolve the social issues currently facing us with cell phones. The use of cell phones violates many social conventions. We are being subjected to hearing the conversations of others in busses, trains, etc. Indeed, many people are opposed to lifting the cell phone ban on airplanes due to the distraction and increased noise level of those making calls [14].

But, in many ways, it is not the call itself that annoys many people. It is rather the attitude that those using their cell phones adopt, either consciously or unconsciously. While using our cell phones many of us adopt an attitude

[104] Recall from Chapter 4 that Opportunistic Communication is communication that happens, only because it is so trivial to do so.

that the social context we were participating in before the call is no longer relevant and is replaced by the social context consisting of us and the remote party. Geren [15] calls this "absent presence", a state where "one is physically present but is absorbed by a technologically mediated world of elsewhere." With their potential for multimodal user interfaces and near effortless communication, mainstream wearable systems could exacerbate this effect.

These issues are magnified if we are participating in a social gathering. If we have been participating in the gathering and suddenly begin talking (and gesturing) to a remote party without any visual cues of that transition, there is likely to be confusion among the others in the group. The others in the group will not have had sufficient visibility into the context switch we have made. As a result, the group may initially think we are still engaged in the group's conversation, although our conversation no longer makes sense within that context. The other members of the group may spend time and cognitive resources trying to reconcile our speech with that conversation. After a while they will realize this is not possible and then assume we are speaking to a person remotely. The expenditure of time and cognitive resources to reach this conclusion and the disruption it caused to the group's conversation could engender ill feelings of the group toward us.

10.1.5 Transparency and Attention in Face To Face Interpersonal Interactions

It is expected that we do not employ hidden agendas when we converse with another person face to face. Failure to follow this convention can make personal meetings awkward and unpleasant and cause us not to trust the other party.

For example, imagine that you know the person with whom you are conversing face to face is recording all or part of your conversation and, in real time, checking what you said for accuracy. It would probably put a pall over the experience, at least as far as you were concerned. You might even be reluctant to meet the person again.

Mainstream wearable systems will give their user the ability to capture the audio and video of a person to person meeting. It will also be possible to query local or remote databases and display information relevant to the conversation, much like the Remembrance Agent [17].

The user of a mainstream wearable will be able to do these activities with little or no visible signs. The other party will never really know if the person is a "passive converser" or an "active converser" who is using wearable technology to try and gain some advantage from the meeting. This could change the dynamics of face to face meetings, as each person tries to maintain an equal level of status in the interaction.

Another issue is remaining attentive to the other person as you receive information from your wearable system. While you are listening to the information in your concealed earpiece or viewing it in your wearable display glasses your ability to effectively converse with the other person may decrease. The other person will notice this as your interaction with them becomes less fluid and occasionally less relevant to what was just said.

In a study done with a wearable system, researchers found that using a heads up display to view information while conversing with another person significantly impaired the quality of the face to face conversation [18]. The impairment was experienced by both the user and non user of the wearable system.

Specifically, users of the wearable system felt that they were listening and concentrating on the conversation less, and paying less attention to the other party when the wearable was active. The party not wearing the system appeared to notice the wearable system user's loss of attention to a lesser extent but still felt that the wearer was not concentrating as much when the wearable display was active.

While a mainstream wearable system will possess the awareness and intelligence to provide received information to the user in the least intrusive and most effective manner for the user's current context, attention impairment during face to face communications could be an issue.

10.1.6 Personal Possessions as Expressions of the Individual

By their nature mainstream wearables will be tightly integrated with their user. This can happen in several ways:

- These devices could be designed into clothing. Techniques such as E-Broidery [19], which embeds circuits into clothing using metallic yarn and gripper snaps, could result in clothes providing built in electronic "buses" for connecting wearable devices.
- Biometric devices are placed against the person's body to monitor a person's health status and take proactive action when abnormal readings are sensed.
- Wearables can be placed in jewelry and watches. We have already seen examples of this. The iButton [20] is a computer chip housed in a variety of form factors, including rings and watches. When the ring is placed against its reader, information stored in the chip is transferred to the reader. The Casio Technowear VDB200B-1 is a GUI based organizer with a touch screen in a wristwatch [21] and Fossil produced a watch with Palm Pilot functionality [22] .

Because of this close association, users will consider wearables their personal possessions. There is evidence that people invest a great deal of their sense of self in the possessions they consider personal [23].

Jewelry is a good example. For many people, their jewelry is a reflection of who they are and what they value. This was made painfully obvious to us during a series of focus groups held to determine people's attitudes toward mainstream wearables. Forty people (20 men and 20 women) from professional and mid management level positions in South Florida were shown computer based scenarios involving the use of mainstream wearables. In one scenario, a woman wore a gold pendant with a red LED embedded in the center (see Figure 10-1). The LED would activate whenever the wearable was used. This would allow those around her to realize that she was communicating with a distant party and not talking to herself, thus addressing the social convention discussed earlier.

Almost without exception, the women liked the concept of the wearable. But they disliked (some strongly!) the glowing pendant. Men had similar feelings but were less forceful in expressing them. Most of the women felt

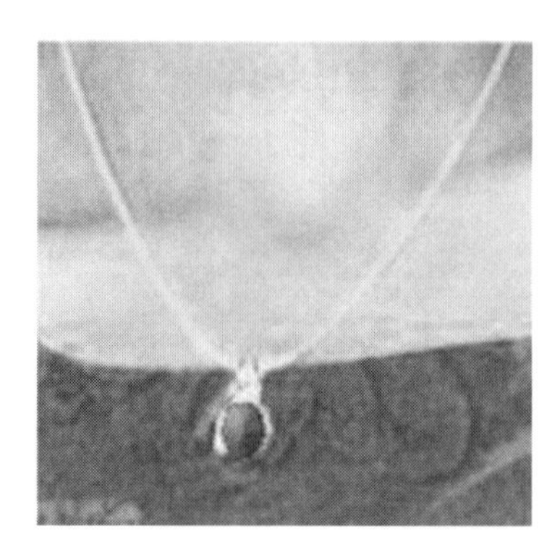

Fig. 10-1. Focus Group Necklace (left) *(Motorola, Inc)*

that their jewelry is very personal and they did not want it “geeked up” with flashing lights.

It is possible that some of this resistance would diminish if the jewelry used for these functions is very elegant and designed by top jewelers, like the MIT Media Lab heart monitor embedded in the upscale diamond studded Heartthrob Brooch [24].

Nevertheless, this is a critical issue for the acceptance of jewelry enhanced with electronics.

10.1.7 The Home as Refuge

Many of the researchers using wearables keep them on most of their waking hours [25]. They wear them at class, at work, and while traveling. Indeed, if wearable devices provide such valuable services, when would we ever want to take them off?

The answer: when we get home. Most of the members of the focus groups mentioned above were very skeptical of the scenarios showing the use of wearables in the home. Many did not think they would ever wear these devices in the home – even though the concept devices shown in the scenarios were very unobtrusive and otherwise appealed to them.

Most focus group members said that when they get home the first thing they do is take off all of the things hanging on their bodies (PDAs, phones, even jewelry), a process we call "technology shedding". They also changed into more casual, comfortable clothes, the type that affords fewer options for attaching devices.

The home is viewed as a refuge, a place to decompress and relax. Most focus group participants felt that wearable technology was incompatible with this relaxed state. This issue may become less serious as mainstream wearable systems start to appear and people see a value to keeping their system with them in the home. This will become especially true as devices in the home become more intelligent and the home becomes a pervasive computing environment which collaborates with the user's wearable system to assist the user with their everyday tasks.

10.1.8 The Value of Self Reliance

Wearable computers can significantly augment our cognitive functions. Always on and always accessible, wearables can offer assistance in almost any situation. Many applications currently being prototyped aim at providing an augmented reality where the wearable system superimposes information over the user's real worldview on the screen of their wearable display.

Most of these applications provide information that supports a task being done. For example, a mechanic can see textual information about a part while repairing an aircraft engine. These systems make an already highly skilled person more efficient and effective.

However, when the application provides detailed, step by step instructions on typical procedures in a task domain, it can have an additional effect. Imagine a repair manual that contains all of the procedures provided step by step using full motion video, high quality sound, and hyperlinked text including detailed, context based help. The user can stop, replay, and zoom in on the information. In addition, he can easily and quickly pose queries to extensive knowledgebase online to resolve almost any difficulty he may face during the task. With such a system, there may be pressure to replace the highly skilled mechanic with a lesser skilled technician. This could have the effect of "dumbing down" the skill level of occupations using these systems.

This is also an issue in performing our everyday tasks. Applications such as the Remembrance Agent and DyPERS seek to assist us in remembering people and experiences [26].

In the South Florida focus groups mentioned above, many subjects reacted with apprehension and concern at the memory assistive applications they saw in the scenarios. They feared that continued, frequent reliance on applications such as reminders and assistance in tracking personal items would reduce them to "mindless robots" being controlled by the technology.

However, self reliance can also be increased with a wearable system. Applications cited in Chapter 3 such as Personal Guardian can make a person feel much safer. Indeed, this may be a strong incentive to buy a wearable system. As precedence consider that almost 90 percent of those surveyed in a University of Michigan study agreed or strongly agreed that the most important reason for having a cell phone was for emergency situations or to let others know when they were running late [26]. The downside of this is that the use of the wearable system and its security and safety enhancing applications may induce some people toward reckless behavior due to an exaggerated sense of personal security [13].

10.1.9 Emotional and Physical Dependency

According to a 2005 poll by Marketing Insight, 23.7 percent of 9,836 respondents refused to part with their hand-held device even for a second, a behavior that could be classified as obsessive [28]. And Europe's first detox clinic for videogame addicts opened in 2006 [29].

Wearable systems, with their ability to provide an immersive virtual reality environment and effortless communication, could pose a significant risk of addiction to those are susceptible to cell phone and video game dependencies. Indeed, it is possible that, given the inclusion of biofeedback, immersive VR environments, and machine learning, the wearable system could foster an emotional dependency that could become a social issue.

Is it possible to get physically addicted to a wearable system? Perhaps. Steve Mann is one of the original cyborgs at MIT. He has been wearing his wearable system almost continuously (while awake) for over 20 years.

While going through a security checkpoint for a Toronto-bound plane at St. John's International Airport in Newfoundland, Mann encountered the reality of increased security after 9/11 [30]. Security guards at the airport told him he had to take off all of his wearable gear and put it through the x-ray machines. He also had to turn his wearable computer off and then on again. Some of the devices were disassembled and visually inspected as well. However, this apparently did not satisfy the security personnel and they allegedly took him to a private room for a strip-search during which, according to Mann, the electrodes were torn from his skin, causing bleeding, and several pieces of equipment were strewn about the room.

He received part of his system back, but the system could not function at its previous level since his glasses containing the embedded computer display were not returned to him. Without a fully functional system, Mann said, he found it difficult to walk and fell at least twice in the airport. He had to board the plane in a wheelchair. He claims he felt dizzy and disoriented.

For several weeks after the incident Mann claimed that he could not concentrate and behaved differently. He underwent tests to determine whether his brain has been affected by the sudden detachment from the technology, but the results were not made public.

This could be a real issue with the acceptance of mainstream wearable systems. The real or even perceived effects on the body from sustained wearing and use of these systems could significantly impede their widespread adoption, even if the effects are significantly less serious than those asserted by Mann.

This story also makes it clear that sensors that contact the body must do so non-invasively. This means easy and painless removal from the body while at the same time adhering to the body sufficiently during use to obtain reliable data.

10.1.10 Remaining In Control

As wearables become more powerful, they will become more proactive. Much of this increased functionality will come from the use of applications that act on information about the user collected from various worn sensors or encountered within the user's Personal Operating Space.

As the number of person oriented, proactive services provided by wearable applications grows, it will become more difficult to know what is being done for us. Many decisions will be made in the name of the user, but without their direct knowledge and control.

This may not be an issue if the decisions involve which web sites to download or deciding where you might want to eat dinner. However, if decisions are being made about the current state of your health or deciding which email messages are kept and which are discarded and you never see, you may want some visibility into what is going on. This visibility is especially important if a sensor malfunctions or an application acts abnormally.

Currently visibility is often provided through direct interaction with the devices that are making the decisions. For example, when I compose a query using a web browser, I know what the search criteria are. If the query is composed autonomously, based upon information gathered without my direct awareness, I will have much less insight into the criteria.

The use of wearables has the potential to magnify this problem. Developments in biosensors, affective computing, and telemedicine will increase the number of decisions being made autonomously that affect us directly. Although most of us in the wearable computing area are comfortable with this and understand the technology, the mainstream population does not understand and is not enamored with the technology. They are simply looking for devices that will make performing their daily tasks easier and more enjoyable. Operational invisibility can often make devices harder to use [31] and raises the anxiety of their users. This loss of control in the devices they use was the greatest concern expressed by the members of the focus groups discussed above.

The loss of control can come from outside the wearable system as well. Computers today are routinely subjected to virus attacks, and sometimes infected, causing serious damage. Bluetooth security vulnerabilities are well known [32]. However, with a wearable system the stakes can be much higher. Your wearable system could be monitoring your health and sending information to a remote health maintenance server or to a physician. Your system could also be launching and controlling semi-autonomous software agents for various complex transactions. Finally, your system could be monitoring the environment for any pathological agents.

If a virus was to infect your wearable, it could compromise any of the above functions with serious results. For example, it could spoof your health maintenance application into sending erroneous biosensor data to your health maintenance company or your physician. This could cause the dispatch of emergency health resources, the cost for which you may be responsible since it was not a real emergency. Even worse, the incorrect data may not trigger any visible response. Instead it may simply be logged in your medical data and result in a distorted view of your health. This could cause increased insurance premiums or even the cancellation of your policy. You would not know about it until it happened, much like a credit report that becomes inaccurate.

If your wearable is used to launch malicious agents or to compromise the performance of those you launched, you could be liable for the damage done and costs of recovery.

Finally, if a virus spoofs your system and those around you into believing it has detected a pathological biological or chemical agent, it could cause a panic and the mobilization of law enforcement and emergency medical responders.[105]

[105] The Department of Homeland Security is developing a system of distributing emergency messages to portable devices including cell phones [33]. It is not clear what security

But the most serious casualty, from your point of view, may be the diminished trust you would place in the performance of your wearable system, on which you have come to depend for providing you with accurate information when needed for your daily tasks.

10.1.11 The Digital Divide Expands

The Digital Divide is a serious problem facing most countries in the world. No matter which side you are on it is an issue of which you need to be aware. “Digital Divide” refers to the gap between those able to benefit from digital technology and those who are not [34].

As wearables systems are more widely adopted their ability-amplifying effect will grow. That is, as more wearable systems are used, more research will be performed leading to more sophisticated and compelling applications which will drive more people to adopt wearable systems, and so on in a virtuous cycle.

This virtuous cycle will cause the digital divide to widen as more and more companies and government agencies seek to create products and services that cater to users of wearable devices. At the same time, fewer products and services will be created for those without wearable devices.[106]

---

measures are being developed to prevent spoofing the system and emitting unauthorized emergency reports.

[106] We saw this process play out for cellular phones. As more and more people adopted cell phones, the phone companies began removing public pay phones. They were a maintenance liability and were no longer profitable since fewer people used them. As a result, it became more difficult for people without cell phones to place or receive a call while mobile.

10.2 RESOLVING THESE ISSUES

If we are to address these issues of the use of mainstream wearable systems, we need to consider the following points:

- Today the dominant use of wearables is in vertical applications such as vehicle maintenance and construction. In the future, most uses of wearables will take place within social settings where the user is in the midst of other people and, most often, in a public place.
- The focus should be on the user's experience. This should not be limited to the interaction with the system. Rather, it must include how the user interacts with and is regarded by others in a social context, even when he is wearing, but not using, the system. We want the wearable system to be as unobtrusive as possible, but still provide others with the visual and audio cues necessary to avoid them believing we are violating social customs and conventions.
- We must address the concerns of wearable users about staying in control of their personal environment and maintaining visibility into the actions taking place that affect them or affect others on the user's behalf. This desire for control of and visibility into the operation of the wearable system will create tension with the goal of minimal interaction complexity and non-use obtrusiveness of the system. For example, continued visibility into the activities of and decisions made by autonomous agents launched by the wearable system may result in receiving notifications when we are busy with some other primary task. This would, strictly speaking, increase the system's non-use obtrusiveness.
- Whenever discussing wearable applications, we should emphasize how the application collaborates with the user. Rather than turning the user into a mindless robot, this collaboration enhances the user's ability to apply his mental energies to more creative and challenging tasks.
- We need to remember that mainstream users are not interested in wearable technology. They do not like visible complexity. They are only interested in those devices that will aid them in performing their everyday tasks. Users will demand comfortable, non-obtrusive, easy to use wearables. This means minimizing the Operational Inertia of the wearable system.

- More research should be conducted on the wearability of wearable devices. The study by Gemperle et. al. [35] is a fine start but more needs to be done if wearables are to truly live up to their name and be used in places like the home.

Failure to address these issues may produce a love – hate relationship between wearable systems and their users, much like the relationship that exists today with cell phones. This was highlighted in a Lemelson-MIT Invention Index study that found that 30 per cent of adults say the cell phone is the invention they most hate but cannot live without [36]. If that happens it will be compelling proof that our wearable systems are anything but transparent.

REFERENCES

[1] S. Mann, WEARABLE COMPUTING as means for PERSONAL EMPOWERMENT, Keynote Address for The First International Conference on Wearable Computing, ICWC-98, May 12-13, Fairfax VA

[2] Becker, B., and Mark G., 1998, Social Conventions in Collaborative Virtual Environments, CVE'98, Manchester 17th-19th June, pp.47-55, available at http://wwwcs.uni-paderborn.de/~bbecker//Becker98.1.pdf, accessed March 26. 2006

[3] I. De Sol Pool, Forecasting the telephone: A Retrospective Technology Assessment. Ablex Publishing Corp., Norwood, NJ, 1983, pp. 129 - 148

[4] Kanellos. M., 2007, Here comes the terabyte hard drive, January 4, 2007, cnetnews.com, http://news.com.com/2100-1041_3-6147409.html

[5] Intille S. S. and Intille A. M., 2003, New Challenges for Privacy Law: Wearable Computers that Create Electronic Digital Diaries, September 15, 2003 MIT House_n Technical Report

[6] Henderson K., 1996, No Dogs Allowed? Federal Policies on Access for Service Animals, Animal Welfare Information Center Newsletter, Summer 1996, Vol. 7 No. 2, http://www.nal.usda.gov/awic/newsletters/v7n2/7n2hende.htm.

[7] Slobogin C., 2002, "Modern studies in privacy law: searching for the meaning of fourth amendment privacy after Kyllo v. United States," Minnesota Law Review, vol. 86, pp. 1393, 2002

[8] Gill v. Hearst Publishing Co. (1953) 40 C2d 224, California Supreme Court, Feb., 17, 1953, http://online.ceb.com/calcases/C2/40C2d224.htm

[9] Prosser W. L., 1960, "Privacy," Califorina Law Review, vol. 48, pp. 383, 1960.

[10] Scheeres J., 2004, Caught With Their Pants Down, Wired News, Nov, 01, 2004, http://www.wired.com/news/culture/0,65502-0.html

[11] Supreme Court of the United States, Wyoming v. Houghton (98-184), 1999. The decision can be read at http://supct.law.cornell.edu/supct/html/98-184.ZS.html

[12] U.S. Supreme Court UNITED STATES V. DI RE , 332 U.S. 581 (1948), available at http://caselaw.lp.findlaw.com/scripts/getcase.pl?court=US&vol=332&invol=581, accessed March 26, 2006

[13] Rosen C., 2004, Our Cell Phones, Ourselves, The New Atlantis, Denville, NJ, Summer 2004, pp. 26-45.

[14] Subcommittee on Aviation Hearing on Cell Phones On Aircraft: Nuisance Or Necessity?, 2005, US House of Representatives Transportation Committee, July 15, 2005

[15] Gergen K. J., Cell Phone Technology and the Challenge of Absent Presence, http://www.swarthmore.edu/SocSci/kgergen1/web/page.phtml?id=manu32&st=manuscripts&hf=1

[16] Tech Digest, 2006, CES 2006: Motorola's tiny Bluetooth headset, available at http://www.techdigest.tv/2006/01/ces_2006_motoro_2.html, accessed March 26, 2006

[17] Rhrodes, B,, 1997, The Proceedings of The First International Symposium on Wearable Computers (ISWC '97), Cambridge, Mass, October 1997, pp. 123-128

[18] McAtamney G. and Parker C., 2006, An Examination of the Effects of a Wearable Display on Informal Face-To-Face Communication, Proceedings of the SIGCHI conference on Human Factors in computing systems, April 22-27, 2006, Montréal, Québec, Canada

[19] Post, R. E. and Orth, M., 1997, Smart Fabric, or Washable Computing, First IEEE International Symposium on Wearable Computers, October 13-14, 1997, Cambridge, Massachusetts. An expanded version is available at http://www.media.mit.edu/~rehmi/cloth/

[20] Dallas Semiconductor Corp., http://www.ibutton.com/index.html

[21] Casio Computer Co., Ltd., http://www.casio.com/timepieces

[22] Fossil Wrist PDA FX2008, 2005, CNET Reviews, http://reviews.cnet.com/Fossil_Wrist_PDA_FX2008/4505-3512_7-31278887.html

[23] Miller, H., 1995 The Social Psychology of Objects, Understanding the Social World Conference, University of Huddersfield,. A summary is at http://www.geocities.com/SoHo/7165/objects.htm

[24] Krantz M., The Ubiquitous Chip, Time.com, December 29, 1997 / January 5, 1998 Vol. 150 No. 28, http://www.time.com/time/moy/daily3.html

[25] Wearable Computing FAQ, http://wearables.www.media.mit.edu/projects/wearables/FAQ/FAQ.txt, accessed March 26, 2006

[26] Jebara T., Schiele B., Oliver N., et.. al., 1998, DyPERS: Dynamic Personal Enhanced Reality System, Proceedings of the 1998 Image Understanding Workshop, Monterrey CA.

[27] PhysOrg, 2006, Cell phone survey shows love-hate relationship, http://www.physorg.com/copyright.php

[28] Tae-gyu K., 2005, Mobile Phone Addiction Emerging as New Problem: Poll, The Korea Times, July 19, 2005

[29] Rauh S., Detox For Video Game Addiction?, CBS News, July 3, 2006, http://www.cbsnews.com/stories/2006/07/03/health/webmd/main1773956.shtml

[30] Guernsey L., 2002, At Airport Gate, a Cyborg Unplugged, New York Times Technology, March 14, 2002

[31] Norman D., 1989, The Design of Everyday Things, Currency Doubleday, New York, pp. 197-198.

[32] The Bunker, Security, Bluetooth, 2004, Security Briefs, http://www.thebunker.net/security/index.htm

[33] Heun C., 2006, Coming Soon To Your Cell Phone: Text Emergency Alerts, InternetWeek, July 20, 2006, http://internetweek.cmp.com/190900447

[34] Digital Divide: What It Is and Why It Matters, DigitalDivide.org, http://www.digitaldivide.org/digitaldivide.html

[35] Gemperle F., Kasabach C., Stivoric J., et.. al., 1998, Design For Wearability, Proceedings of the Second International Conference on Wearable Computing, Pittsburgh, PA.

[36] Cell Phone Edges Alarm Clock As Most Hated Invention, Yet One We Cannot Live Without, 2004, Lemelson-MIT Program, January 21, 2004, http://web.mit.edu/invent/n-pressreleases/n-press-04index.html

Chapter 11

FUTURE OF WEARABLE SYSTEMS

As Yogi Berra said "It's tough to make predictions, especially about the future". Nevertheless, we consider the future of wearable systems in this chapter.[107] Like all other technology, there are several forces affecting the future of wearable systems, including political, legal, social, and technological. In this chapter we assume none of these forces create a climate in which widespread adoption of wearables is precluded.

11.1 TRENDS

If we are to look to the near future (10 years out) of wearables, we can make some relatively safe assumptions:

- **Available processing power and memory capacity will be orders of magnitude greater than available to wearable computers today**.

[107] The secret to prediction, of course, is to predict far enough into the future so that you can remind people when you are right but no one else will remember when you are wrong.

Progress in processor speed has followed Moore's Law since Gordon Moore, cofounder of Intel, observed it in 1965. Moore's Law states that the number of transistors on a chip doubles about every two years [1]. However, many people are not sure the law will continue to hold into the next decade due to the decreasing spacing between the transistors and the increasing power consumption and heat generation produced by the chip.

But chipmakers are developing fabrication techniques to address both issues. Chipmakers say line distances will be falling steadily until they reach 22 nm by 2016 [1], In addition chipmakers are moving toward multicore architectures. By putting multiple processing cores on a single chip, each core can run at a slower speed, consume less power, and produce less heat – and still result in significant increases in overall processing speed.[108]

Another approach is to stack elements of a chip vertically [2]. This shortens the distance between elements, providing more speed without increases in power consumption or heat production. This approach is still experimental but research is continuing on the design tools and fabrication techniques.

- **Component size will continue to shrink**. The techniques discussed above to increase processor speeds of chips can also be applied to pack more functionality onto a chip of the same size. This allows a higher degree of functional integration.

  System on Chip (SoC) architectures are one example. A SoC contains all the elements of a specific functional system [3]. This includes the embedded processor, ASIC logic and analog circuitry, and embedded memory. All of this on a single chip in a much smaller area than if the

[108] The biggest hurdle in the widespread effective use of multicore chips is that software must be designed specifically to utilize the multiple processors. There are currently very few developers that know how to do this and the required techniques are not being widely taught in software engineering or computer science curricula.

separate components were on a printed circuit board. The result is smaller functional components and thus smaller wearable system devices.

- **Speech interfaces will improve and become increasingly adopted**. SoC architectures will allow chips targeted for specific applications [4]. Speech recognizers will certainly benefit from this. Current SoC implementation of speech recognizers support limited vocabulary [5], [6]. However, these systems will become more powerful and will be able to handle larger vocabularies and broader contexts.
- **Component costs will continue to fall**. Wearable systems should ride the cost reduction curves typical of consumer electronics. For example, in 1995 a Seagate 1 Gbyte hard drive sold for $895.00, or $895.00 per Gbyte [7]. In 2007, a 2 Terabyte (2,000 Gigabytes) LaCie Bigger Disk Extreme external hard drive sold for $690 or $0.35 per Gbyte [8]. Complete PC systems have come down in price almost as quickly while at the same time offering significantly more functionality (writable CD ROM. DVD, etc) than their 1995 ancestors (see Table 11-1 below).

Table 11-1. Fall of Prices for Computers and Hard Drives

Year	System	Price	Metric ($/MB-MHz)
	Computer		
1995	120 MHz Pentium processor with 16 Mbyte of RAM,	$2500	$1.3000 $/MB-MHz
2006	3 GHz Pentium 4 processor with 2 Gbyte of RAM	$868	$0.0001 $/MB-MHz
	Hard Drive		
1995	Seagate 1 Gbyte hard drive	$895	$895.00 /Gbyte
2007	LaCie Bigger Disk Extreme external hard drive	$690	$ .35 / Gbyte

This is predicated on a few assumptions:

1. There are no government or regulatory impediments to the widespread adoption of wearable systems;
2. Open standards are adopted for the interfaces between wearable system components;
3. The application execution system (OS, APIs, etc) are standardized and non-proprietary, allowing a vibrant development industry.[109]

11.2 PREDICTIONS

What will the architecture of a wearable system be in 10 years? 25 years? Will we realize the vision of pervasive computing where computing and communication devices are so cheap that there is no economic incentive for exclusive ownership, where every garment contains an instance of our wearable system? Will we still have a hybrid system consisting of a core suite of functions in a small, ZOID device and have specialized devices in our garments or will we have very few devices and application in our wearable system instead leverage off of the ubiquitous smart spaces around us?

Whatever the actual architecture, wearable systems 10+ years from now may have some or all of the aspects discussed below.

[109] It is not clear that the application execution environment needs to be open source, (although that is desirable), for a viable application development community to exist.

11.2.1 Establishment of New Social Conventions

Widespread use of mainstream wearable systems will generate new social conventions and modify current ones. This is a fairly safe prediction as there has been precedence with each new technology.

Some potential conventions society might formally adopt or which may arise informally due to their widespread voluntary adoption include:

- People engaged in a face to face communication will be expected to inhibit their use of recording and verifying mechanisms of their wearable system for the duration of the conversation. Users would be expected to program the recording function to recognize those contexts in which recording was socially unacceptable and have the system inhibit the use of those functions in those contexts.

  If active enforcement proves necessary, there may be regulations that require a visible and/or audible indication that the system is recording and/or verifying something the other party said. This is similar to the audio tone you hear when the remote party on the telephone records your call. It would also be in the same spirit as discussions today about requiring a clearly audible indication that a phone's camera is taking a picture. The idea is to allow the people close to the user to hear the signal and either remove themselves from the situation, confront the camera user, or notify the management of the facility they are in about the improper or undesired use of the camera.

  This convention may extend to people visiting someone's house. Just as cowboys were expected to remove their guns when entering someone's home, users of wearable systems would be expected to inhibit the operation of the video and/or audio recording devices of their wearable system. The audio mechanisms discussed above would ensure that the visitor could not use those facilities surreptitiously.

  Note that the use of the recording device may not even be a deliberate act by the user. If the user's profile indicated a high interest in home architecture or interior design, the wearable may autonomously begin to record when it sees something fitting the profile. The user may not even know this is happening. This illustrates the need for a context based policy of inhibiting the recording function by the wearable itself.

Obviously, these conventions can be temporarily suspended. For example, if a person asks the homeowner if she can take a picture of the bedroom set and the homeowner agreed, the wearable could do so. However, once the current context changes, the wearable would once again inhibit its recording function.

- Schools may enforce bans against bringing wearable systems into exams. The system's ability to read the test question with a concealed camera and understand it using natural language would allow it to form a query and search for the information over the Internet or within the wearable's own data. The information would be summarized and delivered silently to the user over their private audio device.

  Students could prepare queries beforehand requesting help with a test question (ex., 'Need help with question 7') and send them to one another using the wireless short range RF similar to Bluetooth. Small hand gestures detectable by the system but hard to see or understand by the teacher could be preprogrammed to execute these requests silently.

  Enforcing this ban by prohibiting the users wearing their wearable systems in the test area may not be acceptable since the wearable systems may be performing other important duties such as monitoring the student's health, providing medicine, and even prosthetic services. Instead, the school would have to monitor for the use of the wearable systems' long and short range wireless transmissions or jam them.

  Note that this discussion assumes the current educational model of centralized teaching facilities (e.g. schools) is still used. It is possible that within 10 years education would be completely distributed and self driven. In such situations, the student's wearable system could assume the role of instructor or teaching assistant. The wearable system would also ensure that the student took the test honestly.

- The wearable could be legally defined as part of the person. Under this scenario, any attempt to remove it from the user, intercept its transmissions, or search its contents would be regarded in the same way as removing the person's clothes, wiretapping their conversations, and searching their wallet. Under those associations, all of the actions involving the wearable would require a duly executed search warrant.[110]
- Wearables will routinely employ highly intelligent software agents that will carry out sophisticated tasks for the user. These agents can be dispatched by the wearable system itself, often without the explicit knowledge of the user. The agents would traverse the Internet and interact with a variety of other agents in a complex series of transactions that can span a large range of time and space.

  Conventions, both civil and criminal, could arise that hold the user of the wearable dispatching the agents responsible for all effects of the agents activity, whether foreseen by the user or not. Agent programs would be required to log and transmit back a record of all its activities to the user's wearable and to identify its owner to other agents with which it interacts. This allows a level of accountability back to the agent's owner. Tasks may be defined where the agent must request permission of the user (not just the user's wearable) before the task is undertaken.

11.2.2 Wearables Used within Multiple Pervasive Computing Environments

Within 10 years many spaces will incorporate significant computing and communications capability. These pervasive computing environments (PCE)

[110] The opinions and conclusions expressed herein are solely those of the author and not those of any entity with whom the author is or has been affiliated. The author is not a lawyer and has had no legal training, and the conclusions and opinions regarding legal issues should not be taken as legal advice

will be intelligent and capable of semi-autonomous behavior. Examples of PCEs include the home, office, outdoor urban areas, and schools.

Wearable systems will be used within multiple PCEs. The wearable system and PCE will each leverage the knowledge and abilities of the other to amplify the capabilities of the user. In some cases the wearable system will leverage off of the sensing and monitoring capabilities of the PCE to increase its awareness of the user's activities. The wearable will provide elements of the PCE with information about the user such as biometric data and other information about the user that the PCE would have trouble obtaining on its own. The PCE will use this information to tailor its environment to the needs of the user, such as providing the appropriate environmental conditions (lighting, temperature, virtual external views, even reconfiguring the actual physical space as desired or required by the user.

With this frequent and extensive collaboration of the wearable system with elements of PCEs, more frequent communication will take place between the user and his environments than between the user and remote people or machines.

Within 10 years robots will be in many homes and in businesses. Wearable systems will interact with domestic and business oriented robots. The wearable system will provide the robot with the user's information such as email, schedule, and the weather. The robot will use this information to update its understanding of the user's context, needs, and preferences. For example, the robot detects that it is going to rain today. It makes sure the umbrella is by the door so that later, when the user does go out, it is right where she can grab it (since at that time the robot may be doing other things). To facilitate this, wearables and robots will communicate using a machine oriented language. Even if the robots interact with humans via speech – the wearable and robot will not when interacting with each other.

11.2.3 Wearable systems will incorporate personalities

What is a personality? Simon [9] defines it as:

> "The complex of characteristics that distinguishes an individual or a nation or group; especially: the totality of an individual's behavioral and emotional characteristics"

There is some debate as to what actually constitutes a personality. Some psychologists believe that traits are only a convenient device for assigning labels to human behavior while others maintain they are real, internal characteristics that distinguish one individual from another [10].

Meehl [11] categorizes those trait behaviors we can see and label as surface traits. Surface traits describe behavior. He categorizes internal characteristics that presumably direct behavior as source traits. Source traits can only be inferred from observed or reported behavior. They are used to explain a person's behavior.

Psychologists refer to source traits to explain the evolution of traits over a person's lifetime and attribute certain behaviors to motives or needs. Source traits may remain fairly consistent but surface traits may evolve, as new behaviors are acquired and old ones change.

There are several framework models of personalities. One of the most common is the Five Factor Model (FFM) [12], [13]. The FFM defines five personality traits:

- Extroversion: This includes characteristics such as excitability, sociability, talkativeness, assertiveness, and high amounts of emotional expressiveness.
- Agreeableness: This includes attributes such as trust, altruism, kindness, and affection.
- Conscientiousness: this includes high levels of thoughtfulness, with good impulse control and goal-directed behaviors.
- Neuroticism: this includes emotional instability, anxiety, moodiness, irritability, and sadness.
- Openness: This includes imagination and insight, a broad range of interests.

Another model, the PAD (Pleasure, Arousability, Dominance) model of personality and temperament defines a three-dimensional temperament space using nearly independent temperament traits of Pleasure-displeasure,

Trait Arousability, and Trait Dominance-submissiveness [14]. Mehrabian showed how the PAD model maps to the Big Five Framework [15].

There is a lot of past and current research on the application of personality to human computer interaction [16], [17], [18] and in robotics [19]. We know that people's reactions to computers are affected by how the computer responds to them. In The Media Equation [20] Reeves and Nass showed that people react to computers much the same way they react to interactions with humans. For example, people reacted to a computer more favorably when the computer assumed a personality similar to theirs.[111] Another interesting finding is that people reacted even more favorably when the computer seemed to change its personality to one that was like the users.[112]

What goes into expressing a personality? Reevs and Nass showed that it is effective to provide minimal cues such as language style, level of confidence in the computer's responses, and who initiated the dialog to express dominant and submissive personalitics.

However, a mainstream wearable system will interact with the user on a much closer and more frequent level. Therefore, its personality must be deep or it will soon seem superficial and the user will grow tired of it.

The challenge in designing these personalities for wearable systems will be to use the above expressive traits to at least suggest the presence of the different personality attributes defined by the FFM or PAD models.

Speech will be a common means of expressing the wearable's personality using some or all of these expressive speech attributes:

[111] Their study only looked at the personality traits of dominant/submissive.

[112] For example, if users were dominant, they reacted more favorably when the computer changed from a submissive to a dominant personality.

- Prosody. This includes the rate of speech, average pitch, the range of pitch, how pitch changes, vocal intensity, and voice resonance [16].
- Articulation and level of formality
- Idiomatic use of language. This is mostly determined by the culture and subculture. For example, a young urban male will have a different use of idioms than an elderly suburban male.
- Tone. Examples include dominance, submissiveness, arousal, etc.
- Dialog initiation

Compelling expression of these personalities requires a robust speech user interface. Wearable systems in ten years will most likely have the processing power, memory, and algorithms for real time large vocabulary, natural language understanding speech interfaces. This is crucial for accurate rendering of personalities by speech.

Non-speech mechanisms including posture, gestures, and facial expressions can also convey personality traits. These mechanisms could be displayed by a graphical avatar the user wishes to associate with the wearable system.

Users will be able to purchase personas – personalities of famous people such as movie stars, singers, even politicians. In that case, most of these attributes will be determined by the persona. There will also be tools that people can use to develop their own personas for their wearable system.

As the user grows the wearable system's personality will have to evolve and change to reflect a companion of the same (or older) age. This personality evolution profile must take into account the user's personality to ensure the evolution does not produce an incompatible personality. Of course, the user can explicitly select a new persona for the wearable system, thereby installing a new personality.

Why is the design of a personality so important for a mainstream wearable system? The wearable's personality will be an important factor in the transparency of the system. If the user finds the personality inconsistent or incompatible (for example the user is an adult but the wearable system still has the persona of an adolescent), the system may start to annoy or even anger the user and could significantly reduce its transparency.

11.2.4 Wearable Systems Will Be Capable Of Autonomous Behavior

The wearable system will incorporate learning to better understand the user's need and activities and to enable it to be more proactive in assisting the user. Much of the input to the learning algorithms will come from the user context information. They will also use common sense reasoning and user context to flexibly deal with temporal and nondeterministic tasks and situations.[113]

Utilizing its understanding of the user, the context information it collects, and its learning ability, the wearable system will be capable of independent reasoning and goal directed behavior. This will allow it to act semi-autonomously, making both proactive and reactive decisions without the explicit intervention of the user.

This will require both behavior autonomy and goal autonomy [23]. Behavior autonomy allows the system to execute the actions required to reach its goal. Goal autonomy enables the system to reason and decide by itself how to select the next goal and what actions it should take so that its goal is achieved. This means the system will have goal planning capability. We discussed planning in Chapter 6 for context awareness.

This planning system will execute four tasks to provide goal autonomy [23]:

1. Perceive: The wearable continuously receives information about the user and its environment and detects when the context has changed.

[113] See for example LifeNet [21] and EventNet [22], two programs utilizing commonsense reasoning to reason about typical daily tasks and temporally related tasks, respectively.

2. Reason for goal selection: Using its context information, world knowledge, and perhaps commonsense reasoning the planning system infers the next goal.
3. Reason for action selection: The planning system then selects those actions that will have the highest probability of achieving the goal.
4. Act: The selected actions are executed. Behavior autonomy allows the system to take the selected actions without explicit user intervention or approval.

Part of the actions the wearable may undertake to achieve its goals will include the spawning of software agents. These agents will traverse the Internet to gather information, execute transactions on behalf of the user, and negotiate and collaborate with other agents. A major enabler of this capability would be adoption of the Semantic Web [24].

As discussed in Chapter 10, the ability of the wearable system to act semi or completely autonomously and to spawn agents that can act likewise poses unique legal and social issues. It is possible for the wearable system to act in a way that could harm the user; for example, presenting audible information at a volume level that damages the user's ears. In addition, the agents spawned by the system (without explicit user intervention or knowledge) could malfunction and cause financial or even physical damage. An important design issue is providing traceability of the agent's actions and setting conditions for which explicit consent of the user is required for the agent's actions.

In addition, the behavior will have to be constrained by rules or laws of conduct established through legal and social mechanisms. While there is some research on the legal liability of software agent designers [25], [26], the use of software agents is still too recent to provide much guidance. The legal case law developing around computer viruses [27] may provide insight into the legal issues for autonomous agents, since many computer viruses exhibit this behavior. An example of rules that could inhibit detrimental behavior of a wearable system is the Laws of Wearables discussed later in this chapter.

Fig. 11-1. A Tamagotchi

11.2.5 People Will Develop Long Term, Symbiotic Relationships with Their Wearable Systems

People will develop long term, symbiotic relationships with the core wearable system. They will come to rely on the wearable, much like some people depend on butlers or nannies.

There is precedence for people bonding with virtual entities. In the late 1990s children became very attached to Tamagotchi pets (see Figure 11-1). Children (and some adults) suffered emotional trauma when they neglected to 'feed' their Tamagotchi (by pressing buttons on the case) and it 'died'. One woman ran over a biker in a frantic attempt to get to her Tamagotchi and 'feed' it [28]. There are even Tamagotchi cemeteries (both online and physical) where children brought their Tamagotchis after they had 'died' [29].

Whereas children used to attribute life to objects that moved under their own power, they are now starting to perceive psychological rather than physical properties as criteria for life [31]; or as Turkle puts it "...motion gave way to emotion and physics gave way to psychology as criteria for aliveness." [31]. Children are broadening the criteria for what it means to be alive and are forming relationships with electronic devices that meet that criterion.

In ten years many of these children will be young adults and forming a relationship with a wearable system that provides them constant assistance, understands their preferences and activity patterns, and exhibits a multi-faceted personality when interacting with them will seem quite natural.

All of these predictions assume wearable design accommodates evolving social concerns and customs. Failure to do so could significantly change, delay, or even prevent some of these developments from occurring.

11.3 LAWS OF WEARABLES

In 1940 Isaac Asimov stated the "Laws of Robotics" [32]. To these three laws he added another ("The Zeroth Law") in 1985 [33]. In his books these laws were created by society to constrain the actions of robots, which were capable of independent, proactive behavior.

In a similar way, as wearables become more intelligent and capable of proactive behavior, it will become necessary to constrain their actions. The objective is to prevent, or at least minimize, the possibility of a wearable causing harm to a person, even at the behest of its user.

In this spirit, and with apologies to Dr. Asimov, here are the "5 Laws of Wearable Systems" (5LoWS).

1. **A wearable may not harm its user, nor cause its user significant discomfort.** This means the wearable must be aware of the user's situational context. For example, the device may increase the volume of its responses in order to compensate for high ambient noise. However, it must not increase it to the point of causing ear damage to the user.
2. **A wearable may not cause harm to any other person or wearable system on the behalf of its user.** With the use of intelligent software agents and Internet connections, it is possible for a malicious user to use their wearable to attack another user's wearable system or data or, through negligence on the part of the agent's designer, allow the wearable to harm someone. The wearable must monitor the effect of its actions and have the capability to refuse commands that cause harm.

3. **A wearable must carry out the requests of the user, but do so in a manner that does not conflict with the previous laws**. The wearable system must give priority to those commands that enable the completion of the user's primary task, even if it means canceling or delaying the completion of support or device tasks. However, it cannot cancel these tasks or suspend these tasks if doing so would harm the user or someone else, even if it means delaying or refusing to carry out the user's current command.
4. **A wearable must deliver information destined for the user in a timely fashion and present it to the user in a manner that does not conflict with the previous laws**. For many of the tasks the wearable will be doing, either by direct request or proactively, time will matter. The system must try to ensure that the command is carried out in a timeframe that makes its results relevant to the user. However, it can't do that at the expense of harming someone.
5. **A wearable must operate as efficiently as possible as long as such operation does not conflict with the previous laws**. Since the wearable system will be operational most of the user's waking hours, efficiency of operation is important. But efficiency must take a back seat to the safety of its user and other people.

Asimov's laws of robotics were primarily a literary mechanism rather than a formal work of cyborg psychology. Like those laws, the above laws of wearables did not develop from any formal study into human – cyborg relations or from deep insight into cyber psychology nor have they been validated. However, they may be a good vehicle to start a debate along these lines.

11.4 CYBERTWIN: A POSSIBLE FUTURE

Let us now gaze into the more distant future to see a mainstream wearable system at its full potential. By then the wearable system is a lifelong companion specifically customized for its user - a CyberTwin. Let's follow the life of a person fortunate enough to have a CyberTwin.

New Chicago, 2163. So, you want to know how CyberTwins were used? Let me tell you about my grandmother and how she used hers.

The Beginning

Each person would get their CyberTwin at birth. It would initially be programmed with the newborn's vital medical and genetic information. This information would be used to help the CyberTwin monitor the health of the infant:

My grandmother, Alice, was born on July 17, 2031. Unknown to her at the time, one of the first things her parents did upon bringing her home was purchase a CyberTwin for her. The hospital provided her parents with a memory cube containing all of her genetic information as well as current biometrics. Her parents gave the memory cube to the CyberTwin technician who programmed the CyberTwin with this information and the profiles of both parents. Also added was basic social knowledge and child psychology and learning techniques. Once programmed, the CyberTwin became permanently bound to Alice. Even if it was stolen, its intimate connection with her formed by this information rendered it unusable to anyone else. It was to be her constant companion for all of her expected 117 years of life.

Alice knew none of this, of course. Nor did she immediately understand the significance of the device that seemed to be always with her. Her CyberTwin, however, understood its purpose and began observing its surroundings and Alice's interactions within it. The CyberTwin continually monitored Alice's vital signs and health via the sensors embedded in her garment. For Alice's parents this was a real comfort, since there had not been a single case of Sudden Infant Death since the monitoring system was developed in the second decade of the 21st century.

The Early Years: The Climb

During the early years, the CyberTwin's primary task would be to help the youngster develop language and to stimulate its curiosity by providing mental stimulation. It would monitor the child's activity to learn the behaviors, preferences, and habits of its user. These would be used to build a profile of the user, allowing CyberTwin to eventually be proactive:

As Alice grew through early childhood her CyberTwin's primary task was to help her develop language and to stimulate her curiosity by providing mental stimulation. The system played music, told stories, and provided simple developmental lessons. The CyberTwin did not replace or diminish the role of Alice's parents in raising her. Quite the contrary, it provided resources her parents used to enhance their interactions with Alice and increased the effectiveness of their parenting.

Alice's parents had made it a point to purchase the latest in smart appliances and home intelligence devices and applications. As a result, their house was a pervasive computing environment. Alice's CyberTwin interfaced with elements of this environment, leveraging the environment's intelligence and monitoring capabilities to learn Alice's activity patterns and implicitly defined preferences. It used this knowledge to update and expand its understanding of Alice's current and probable next actions and thus assisted her and helped ensure her safety. All of this information was stored in the CyberTwin's ever growing behavioral profile of Alice.

As Alice began to crawl and then walk, her CyberTwin monitored her actions to ensure Alice's safety. Its database provided a large number of facts about Alice's home, her parent's activities, and typical living patterns. This allowed CyberTwin to make reasonably intelligent inferences for proactive decisions and to notify her parents if necessary.

When Alice began talking, CyberTwin's special child speech development application and its ability to learn and track the development of Alice's speech enabled it to dynamically create and modify speech recognition vocabularies and grammars to actively help Alice in her speech development.

Adolescence and Young Adulthood: The Summit

Through adolescence and young adulthood the CyberTwin would help with socialization, managing the user's developing social networks, and aid the user in adopting social responsibilities. As a result, the user develops a deep relationship built upon the close companionship provided by the system:

Like all children her age at that time, adolescence was a turbulent period for my grandmother. Her CyberTwin regularly consulted an online social counseling service, to which her parents had subscribed, to ensure their daughter's CyberTwin had the latest and best information about adolescent issues and social habits. It facilitated Alice learning social customs and aided in social collaboration and community formation. Her CyberTwin became Alice's sounding board for her frustrations, excitement with dating, problems with cliques, and other experiences. Since CyberTwin recorded these interactions between it and Alice, it became her diary as well.

Alice's CyberTwin became an active assistant in maintaining and expanding her social network. It constantly searched the Internets (there were three of them at that time) for new social communities and helped Alice join them. It also acted as her interlocutor with her current social networks to minimize their disruption to Alice's current task.

At this time Alice began to develop a symbiotic relationship with her CyberTwin. She changed the name from 'CyberTwin' (the default name) to Abrina, after one of her favorite singers. (She originally chose Kael, after a dreamy movie star. However, her boyfriend objected, saying he did not want her 'being with that guy' when they were together. The boyfriend soon became history.). She also purchased and downloaded the voice and dialog style module of that singer. Now her CyberTwin sounded and spoke just like the singer Abrina and Alice felt that she was actually friends with her. She had also downloaded the singer's avatar which was used as the virtual embodiment of her CyberTwin in her system's displays.

When my grandmother became 16, she decided to get a MindPort. Her parents took her to a neuro-psychologist who ensured that she was ready. Her MindPort was a subcutaneous implant in the base of the neck that allowed her CyberTwin to provide sensory input directly to her brain and receive thought based input from her mind. In the early 21st century these were known as Brain – Computer Interfaces and were much more primitive. With her MindPort she was able to interact with her CyberTwin much more efficiently for many applications.

Adulthood, the Long Coast

During the user's middle age the CyberTwin extends and enhances the user's innate capabilities. Its objective is to empower the user and increase the user's quality of life. It also administers the minutia of the user's life:

After graduating college with a PhD in cyber-psychology Alice went to work for a large CyberTwin manufacturer. She advanced to head a department in cyber behavior management and 5LoWS compliance.[114]

Throughout her career, her own CyberTwin (which she again renamed and re-personalized after getting her job) worked closely with her to assist her and extend her capabilities. It managed her work flow and appointments, working directly and negotiating with other CyberTwins and software agents on the old Semantic Web.

Alice enjoyed being home with her family and used the E-Conferencing system of her CyberTwin to minimize her need to travel. Using the terabit per second speed of the Internet4, then the fastest of the tiered Internets, the meeting participants' CyberTwins collaborated among themselves to recreate almost entirely the experience of attending a meeting in person, using immersive Virtual Reality, biosensor transmission and feedback, and group dynamics tracking. More and more meetings used E-Conferencing after the nanobot bomb scare made air travel so difficult.

Despite her hectic work and family schedule, Alice kept up on her entertainment interests. Her CyberTwin regularly dispatched intelligent, autonomous search bots which searched the Internets for new events and sites that conformed to her interests and were compatible with her schedules.

[114] 5LoWS compliance was the heart of advanced CyberTwin performance, starting in the 2040s and 2050s. It ensures compliance with the Five Laws of Wearables (5LoW) composed in the first decade of the 21st century. The compliance system is informally referred to as a 'cyber-conscious'.

In this way Alice was frequently pleasantly surprised by the new opportunities her CyberTwin suggested to her.

About this time (the 2080s) the debate about cyber-consciousness began in earnest. With CyberTwins and domestic robots acquiring personalities, semi, and eventually completely autonomous behavior, and the ability to converse using completely natural language, many people began to assert that these entities were thinking and conscious. These entities routinely passed the Turing Test, an exercise that people in the 20th century believed was the determining test of whether a machine was thinking or not. (Of course we now know this test was inadequate for this purpose for many reasons).

My grandmother had already come to regard her CyberTwin as another thinking entity. She would often tell us about what Hector (her final CyperTwin's persona) did or found for her or the discussions she would have with him (she always now referred to her CyberTwin as a 'him', not an 'it'). She never said to me if she actually believed it was a conscious entity. In any event, by 2115, this debate was largely over and these entities were afforded some legal rights.

Although Alice visited her family doctor only once every 2 – 3 years, her health never suffered. She had a virtual visit with the doctor (actually, the doctor's avatar, but it looked and spoke just like him) every 6 months. The doctor was up to date on the state of Alice's health since her CyberTwin regularly sent him health information about Alice. This allowed the doctor to examine her online in real time via her worn biosensors and devices such as weight scanners, blood pressure imagers, and thermometers, all of which sent their data to Alice's CyberTwin which relayed it to the doctor. If required, the doctor would prescribe medicine during the virtual visit. If there was a problem with one of the medicines, the doctor was notified immediately and changed the prescription.

The Elder Years: The Decent Begins

As the user ages, the CyberTwin begins to focus on compensating for the user's diminishing cognitive and memory capabilities. More emphasis is placed on monitoring the user's health and gently reminding the user of required tasks and behaviors:

As Alice aged, her CyberTwin began to focus on compensating for her diminishing cognitive and memory capabilities. Her CyberTwin, in collaboration with the sensing infrastructure of her home, ensured that she did not get into any dangerous situations. For example, it turned the lights on to light her path (and turned them off to save energy costs), anticipated her use of appliances and prepared them as well as ensuring they were turned off after being used.

Her CyberTwin reminded her to take her non-implanted medicines and monitored her compliance, sending messages of non-compliance to her doctor and her two children that lived close to her so they could intervene if necessary. It was very accurate in its monitoring since it communicated directly with the intelligent medicine dispensing mechanisms to detect non-compliance. Now, of course, this is no longer necessary. Any medical condition not mitigated through genetic therapy is treated with medicine dispensed via subcutaneous implants.

Alice always liked walking, even if it was just to the store. She never used a Personal Transport Prosthesis. However, when she became 81 and approached the end of middle age, she started to become forgetful. Her CyberTwin had learned the routes she took for her walks and errands. It monitored her path and, if she started to significantly deviate from it or started to wander, it helped to return her to the route. Failing that, her CyberTwin alerted her nearby son (my father) that she was deviating. His CyberTwin passed the alert along to him in the most effective way possible given his current situation to ensure he got the message. He then came over or his and Alice's CyberTwins facilitated him helping her remotely.

The Last Years: The Final Decline

During the final years CyberTwin aids the user in performing as many Activities of Daily Living as possible. It draws heavily on the sensing and monitoring infrastructure of the user's space. It also becomes a cognitive prosthesis to reduce the effects of any major cognitive impairment. And it seeks to maintain the user's social networking:

At 116 my grandmother developed dementia. However, her CyberTwin mitigated many of the effects of the disease and Alice remained active and occupied with the activities of daily living. Although she had forgotten how to use some of the devices in her house's multimedia center, her CyberTwin controlled them for her. Upon her request, it replayed experiential recordings of past events in her life, events with her late husband, her children, us grandchildren, and friends. The experiential recordings provided full sensory feedback, audio and video to be sure, but also smell (and thus taste) and touch. The video was sent directly to her visual cortex via her MindPort, bypassing her impaired eyes. The audio was sent directly to her artificial cochlea.

Alice's CyberTwin also allowed her to interact more fully with her children and grandchildren when we came to visit. Using a wearable micro camera her CyberTwin recognized the faces of her visitors and discreetly told her who they were. Of course, some of us, especially our parents and us older grandchildren knew this, but it helped us feel she recognized us. As we discussed events, our own CyberTwins sent images to hers so she could experience them with us.

Eventually Alice could no longer get out and see her surviving friends. However, she still got together virtually with several of them every week. Her CyberTwin organized and setup the E-Conferencing meeting in collaboration with her friend's CyberTwins. The system was enhanced in 2127 to provide total immersion using Alice's MindPort and sensory feedback. This continued to give Alice the connectedness she needed.

The End. Or Not

Upon the user's death, the CyberTwin could be recycled or used by the surviving family members to create a continuing sense of presence of the user:

After a long and full life, my grandmother succumbed on November 2, 2152. At 121, her cranial integrity could no longer support nanobot neural reconditioning. My grandmother had prepared for her eventual death. She had made out her legal will, her living will, and her CyberTwin disposition document.

Alice was not one for long goodbyes and so had planned on the 'redeployment option' for her CyberTwin. Under this option her CyberTwin would be recycled. Its data and programming would be erased and replaced with that of another newborn. Then the process would start over. All knowledge and memory of Alice in the CyberTwin would vanish.

However, several of us had become very close with grandma and had told her it would break our hearts to think we would never see her again. So she chose the 'sustained memory' disposition option for her CyberTwin. Now her CyberTwin drives a virtual presence system that creates a continuing sense of presence of our grandmother for us.

The system utilizes its vast knowledge of Alice in her CyberTwin to answer questions put to it the way she would, complete with her dry humor and quick wit to which we were subjected so often. It displays her actual image with lip synced animation, head movement, and speech prosody characteristic of Alice as it responds. Each system utilizes a free space, life size holographic projector to provide a compellingly lifelike image. By utilizing the high speed Internet3, Alice's CyberTwin drives several of these systems independently, creating the sense of a separate presence for many of us despite our geographical dispersion.

Eventually, of course, once all we grandchildren have died, Alice's CyberTwin will probably be recycled.

Had my grandmother died much sooner, my parents say they could have had aging algorithms age Alice's image in the virtual presence system over time to simulate her continued aging in her virtual life as we aged as well. My mother knew a family who lost a child in an accident that did this.

Epilogue

My grandmother spent almost 121 years with her CyberTwin. In that time, she grew to regard it as her constant companion, a sort of electronic twin (hence the name CyberTwin I guess). At times it was her best friend, especially during the sometimes lonely elder years. And now, as a final service, her CyberTwin is giving us the gift of her continued presence – if only virtually.

REFERENCES

[1] Moore's Law, 2006, Intel Technology and Research, http://www.intel.com/technology/silicon/mooreslaw/

[2] Aston A., 2005, More Life For Moore's Law, BusinessWeekOnline, June 20, 2005, http://www.businessweek.com/magazine/content/05_25/b3938629.htm

[3] Shini M., 2005, System On Chip – SoC, JTAG course, http://www.cs.huji.ac.il/course/2005/dft/Presentations/Summer%202005/Mohanad%20-%20System%20On%20Chip%20(SOC).ppt

[4] Ohr S., 2003, Speech-recognition core targets high-volume apps, EE Times, http://www.us.design-reuse.com/news/news2008.html

[5] Ji T., 2002, DSPs Tackle Speech Recognition for 3G handsets, CommsDesign, file://///HAL/Wearables%20Book/Backup/User%20Interface/Speech/DSPs%20Tackle%20Speech%20Recognition%20for%203G%20Handsets.htm

[6] Microcontroller delivers voice-enabled solution, 2005, ThomasNet Industrial NewsRoom, September 12, 2005, http://news.thomasnet.com/fullstory/466971/24

[7] Historical Notes about the Cost of Hard Drive Storage Space, http://www.alts.net/ns1625/winchest.html

[8] NextTag Comparison Shopping, http://www.nextag.com/2-tb-hard-drive/search-html, accessed July 17, 2007

[9] Simon, H. (1967). Motivational and emotional controls of cognition. Reprinted in Models of Thoughts, Yale University Press, (1979), pages 29-38.

[10] Inscape Publishing, Inc. , 1996, DiSC® Classic and Models of Personality

[11] Meehl, P. E. (l991). In Dante Cicchetti, & William M. Grove, Thinking clearly about psychology. Minneapolis: Univ. of Minnesota Press.

[12] McCrae, R. R., and John, O. P. (1992). An introduction to the five-factor model and its applications. Special Issue: The five-factor model: Issues and applications. Journal of Personality 60: 175-215, 1992.

[13] Van Wagner K., The "Big Five" Personality Dimensions, http://psychology.about.com/od/personalitydevelopment/a/bigfive.htm

[14] Mehrabian A., 2005, A General & Powerful System for Assessing Temperament & Personality, http://www.kaaj.com/psych/scales/temp.htm

[15] Mehrabian, A. (1996). Pleasure-arousal-dominance: A general framework for describing and measuring individual differences in temperament. Current Psychology, vol. 14, pp. 261-292.

[16] Picard, R., 1997, Affective Computing, The MIT Press, Cambridge, MA.

[17] Galvao, A.M., Barros, F.A., Neves, A.M.M., et. al., 2004, Persona-AIML: an architecture for developing chatterbots with personality, Proceedings of the Third International Joint Conference on Autonomous Agents and Multiagent Systems, 2004. AAMAS 2004.

[18] André, E., Klesen, M., Gebhard, P., Allen, S., & Rist, T. (1999). Integrating models of personality and emotions into lifelike characters. Proceedings of the Workshop on Affect in Interactions – Towards a new Generation of Interfaces (pp. 136–149). Siena, Italy.

[19] Suzuki, N., and Katagiri, Y. 2005, "Prosodic alignment in human-computer interaction", CogSci2005 Social Mechanism for Android Science Workshop (Aug. 2005)

[20] Reeves B., Nass C., The Media Equation: How People Treat Computers, Television, and New Media Like Real People and Places, Cambridge University Press, New York, 1996

[21] Morgan, B.; Singh, P., 2005; `Elaborating Sensor Data using Temporal and Spatial Commonsense Reasoning, http://web.media.mit.edu/~neptune/lifenet/morgan-singh-sense_and_sensors-20051127.pdf

[22] Espinosa J, Lieberman H: 2005, EventNet: inferring temporal relations between commonsense events. Proceedings of the Mexican Conference on Artificial Intelligence. LNCS Springer

[23] Shen Z., Gay R., Miao Y., et. al., 2004, Goal Autonomous Agent Architecture, Proceedings of the 28th Annual International Computer Software and Applications Conference

[24] Herman I., 2003, Introduction to the Semantic Web, http://www.w3.org/2003/Talks/1112-BeijingSW-IH/Overview.html

[25] Heckman C., Wobbrack J. O., 1998, Liability for Autonomous Agent Design, Autonomous Agents 98.

[26] Apistola M, Brazier F.M.T., Kubbe O., 2002, Legal aspects of agent technology, Proceedings of the 17th BILETA Annual Conference

[27] de Villiers M., 2004, Computer Viruses And Civil Liability: A Conceptual Framework, Tort Trial & Insurance Practice Law Journal, Fall 2004 (40:1)

[28] Anders L., 2005, Takes a Lickin', But Keeps On Tickin', September 04, 2005, http://www.louanders.com/2005_09_01_archive.html

[29] Tamagotchi Memorial, http://www.tamatalk.com/IB/index.php?s=5ed9cf69d2eea44975b32484e339f140&showforum=23

[30] A special place for Tamagotchi interment, CNN Interactive, http://www.cnn.com/WORLD/9801/18/tamagotchi/index.html

[31] Turkle, S., 2000, Cuddling up to Cyborg babies, The UNESCO Courier, September 2000, http://www.unesco.org/courier/2000_09/uk/connex.htm

[32] Asimov I., 1968, I, Robot (a collection of short stories originally published between 1940 and 1950), Grafton Books, London, 1968

[33] Asimov I., 1985, Robots and Empire, Grafton Books. London

A Final Note

WHY TAME THE BORG?

Mainstream wearable systems will be difficult to make. Their need for transparent use will require a great deal of creativity and effort. Why go through all the trouble? Why not simply continue to design wearable devices the way electronic devices are currently designed, with their high levels of Operational Inertia and less than satisfying experiences? Why work so hard to tame the Borg?

One word. Life.

If we subjugate our experience of life and all it has to offer to the needs and tyranny of our technology, we will truly be on our way to becoming the Borg. We will be assimilated into a world where our first priority is attending to the needs of our electronic devices. No way you say? When was the last time your cell phone rang while you were talking to someone face to face and you did not interrupt your conversation to answer it?

When was the last time you spent many minutes wrestling with Word or Excel or any other software program, trying to make it do your bidding, your original task now a fading memory?

We may already be on our way toward assimilation.

A few years ago there was an ad for a wireless carrier showing a man fly fishing in a pristine stream in the mountains. He is surrounded by tall evergreen trees, the clear water, and blue sky. The ad copy states that, even here, he is connected to his office and can receive calls. The ad was touting the carrier's extensive coverage. But it was also inadvertently pointing out how much we are willing to sacrifice to attend to our technology.

Consider. The phone rings. The man's attention is yanked from the solitude of his fishing to the phone, its loud ringing a jarring intruder into the serenity around him. He places his rod, line and everything else he is holding into one hand. He then reaches into his fishing vest for the phone and takes it out and looks at the caller ID. The office. His mood changes, from one of relaxed ease to attentive tension. He starts the conversation. He may even move his eyes downward, away from the peaceful scene in front of him to better focus on the conversation. He is no longer in the peaceful stream, but in the hectic chaos of his workplace. Once the call is over he struggles to reclaim the calm, relaxed mood of his environment. He pushes the conversation and issues of the call from his mind, but they are ever lurking, struggling for his attention as he sighs and makes a new cast into the stream.

It may be, as some like Ray Kurzweil [1] and Andy Clark [2] have said, that man and machine will eventually merge as we incorporate more and more technology into our bodies and, eventually, our minds. But, if we do, we must do so on our own terms, not those of the technology. We must do it in a way that allows us to continue to experience and appreciate the world and people around us as our first priority.

Wearable technology is advancing quickly: light, wearable displays, shrinking cell phones, small body sensors, tiny implantable devices. All promise to make our life better in some way.

However, unless they become transparent to use, they will increasingly reduce our ability to appreciate the world around us. These devices must remain in the deep background of our attention, popping up only when absolutely necessary and then only briefly as they provide us with the required information as efficiently as possible. The rest of the time they are assisting us in the background, amplifying our ability to experience and appreciate life.

In the final analysis, wearable devices and systems are just tools. They and their technology do not define us. Appreciation of music, art, nature, friendship, defines us. If we do not make our devices transparent to use, we will indeed become the Borg.

And we will have only ourselves to blame.

REFERENCES

[1] Kurzweil, R., 1999, The Age Of Spiritual Machines, Viking Press, New York

[2] Clark, A., 2003, Natural – Born Cyborgs, Oxford University Press, New York

GLOSSARY OF WEARABLE TERMS

Activity Based Design	A design methodology that emphasizes the activities afforded by and performed with the device or service. See http://www.jnd.org/dn.mss/humancentered_desig.html.
Activity Recognition	Recognizing what a person is currently doing by analyzing data from relevant sensors and an understanding of the user's goals.
Affective Computing	Computing devices and services that can recognize, understand, and express emotions at some level.
Assistive Technology	Commonly refers to "...products, devices or equipment, whether acquired commercially, modified or customized, that are used to maintain, increase or improve the functional capabilities of individuals with disabilities..." Assistive Technology Act of 1998.

Audio Icon	Auditory icons that map objects and events in the interface onto everyday sounds that are reminiscent or conceptually related to the objects and events they represent.
Audio Interface	A user interface that employs audio for input and output. Speech is typically not a large part of an audio interface since it is included more effectively in a speech interface.
Augmented Reality	Augmentation of human perception by overlaying computer generated or enhanced information.
Barge-In	Interrupting the output of a speech synthesizer by speaking. The speech interface detects speech from the user and stops the speech synthesizer output so the speech recognizer can recognize what the user is saying.
Beat Gestures	Hand and arm gestures done, sometimes unconsciously, to emphasize a spoken point.
Behavior Autonomy	The ability of a wearable system to determine its response to received information without direct control of the user.
Biosensor	A sensor that monitors a specific body function.
Body Area Network	A wireless network with a range of no more than 2 meters primarily used to connect devices worn on the body.

Cell Tower Triangulation	Using the signal levels from three cell towers to determine a device's location.
Cognitive Assistance	Assisting the user in a way that compensates for impaired cognitive abilities, whether the impairment is permanent or situational.
Cognitive Load	A measure of how much attention and mental effort the user must apply when using a service or device.
Cognitive Load Theory	"Describes how the architecture of cognition has specific implications for the design of instruction". – Wikipedia
Command And Control	The speech interface recognizes the verbal equivalent of mouse and menu commands or a string of mouse commands. Other commands may be recognized, including changing applications and turning the recognizer off.
Concept Generation	In humans, the act of conceiving a thought to be expressed. In a wearable system, invoking a specific algorithm that results in the need to send a message to the user.
Context Awareness	Any information that can be used to characterize the situation of an entity. An entity is a person, place, or object that is considered relevant to the interaction between a user and an application, including the user and applications themselves (from Dey and Abowd).

Conversational Speech Interfaces	A speech user interface that supports natural language dialog between the user and the wearable system.
Cybertwin	A mainstream wearable system at its full potential. The wearable system is a lifelong companion specifically customized for its user.
Cyborg	A human being with bodily functions aided or controlled by technological devices. For more information see http://whatis.techtarget.com/definition/0,,sid9_gci296606,00.html.
Data Fusion	The merging of multiple pieces of low level data into a high level, easy to comprehend piece of information.
Deictic Gestures	Gestures that refer to the space between the user and those interacting with him. An example is pointing to a spot on the floor where someone should be standing to make a point.
Device Operational Inertia	**Operational Inertia** possessed by a device.
Device Task	Tasks relevant only to the operation of the device or service. The user sees little or no relevance to their primary task.
Dialog	An activity characterized by one or more bi-directional interactions between the user and the device within a specific context.

Digital Divide	The gap between those able to benefit from digital technology and those who cannot.
Disfluency	An interruption in the smooth flow of speech by a pause or the repetition of a word or syllable (for example, "uh", "oh", "um".
Dissaggregation	The distribution of elements of an integrated device among several, collaborating devices. A common disaggregation scenario is distributing the display and input functions of a cell phone to other devices, leaving only the RF functions in the phone.
Divided Attention Problem	The difficulty a person has of attending to multiple stimuli at once. Divided attention is an issue with wearables since they are used in support of other tasks the user is doing or concentrating on.
E-Broidery	See **Electronic Embroidery**.
E-Clothing	Clothing with attached or embedded electronics.
Earcon	Short segments of musical tones, originally developed to provide audio feedback of GUI actions.

eGarments	See **E-Clothing**.
Electronic Embroidery	The patterning of conductive textiles by sewing or weaving processes to create computationally active textiles. Examples are fabricating electronic circuitry on wash-and-wear textile substrates.
Emblematic Gestures	Gestures typically forming symbols that represent specific concepts; for example, the circle formed by the thumb and middle finger to represent 'OK'.
Example Based Help	Providing help to the user with examples rather than detailed explanations of steps.
Eye Tracking	Mechanisms to detect and track the movement and position of the eye to determine gaze direction or movement patterns.
Five Factor Model	A model of personality consisting of five general personality traits: Extroversion, Agreeableness, Conscientiousness, Neuroticism, and Openness.
Galvanic Skin Response	A method of measuring the electrical resistance of the skin. It is highly indicative of some kinds of emotions in some people. Fear, anger, and startle response, are all among the emotions which may produce similar GSR responses.
Gesture Contour	The characteristics of a gesture that identify it to a gesture recognizer. These include shape, speed, amplitude, etc.

Gesture Interface	An interface that uses gestures to specify commands.
Gesture Thresholds	The minimum amplitude of the parts of a gesture contour that will indicate to the wearable that the gesture is an interface command.
Gesture Vocabulary	The set of gestures recognized by a gesture interface.
Goal Autonomy	The ability of the wearable system to reason and decide by itself how to select the next goal and what actions it should take to achieve that goal.
Graphical User Interface	An interface employing graphical elements enabling the user to give commands to the wearable system with a minimum of text input. Most GUIs employ the WIMP paradigm. See **WIMP**.
Gulf Of Execution	The span between how the user expects a system to act and how the system actually acts.
Haptics Device	A device that provides a direct mapping or a translation of real or computerized visual or audio information into tactile stimulation.
Haptics Display	Displays that present tactile renderings of concepts that are not a translation of visual or audio information.

Haptic Interface	An interface that uses tactile stimulation to specify commands and data to the wearable system.
Haptics	The science that deals with the sense and perception of touch.
Head Mounted Display	See **Head Worn Display**.
Head Worn Display	A personal display worn on the head.
Iconic Gesture	A gesture illustrating features in events and actions, or how they are carried out. An example is mimicking the movements of an action such as typing on a wrist worn keyboard.
Immersive Display	A display in which the user's view of the real world is totally replaced by information created by the computer. Immersive displays completely cover the user's eyes and are often used with video games. They are normally not suitable for wearable systems since the wearable system is meant to augment and not replace the user's vision.
Interaction Complexity	The measure of how hard a device is to user for its intended purpose once it is setup and is ready for use. This includes how hard it is to remember and execute commands, the difficulty in getting help, etc. Interaction Complexity is one element of **Operational Inertia**.

Inherent Complexity	The complexity of a device that remains when it has been made as simple as possible. Inherent Complexity is due to the very nature of the device. A device cannot be made less complex without changing one or more of the basic technologies upon which it depends. Contrast this with **Visible Complexity**.
Interaction Mode	The different ways in which a wearable system and user interact to exchange information between each other and the environment.
Intent Recognition	Recognizing the reason why the user performs a task.
Intent Understanding	Understanding the reason why the user performs a task.
Interface Personalization	Techniques for personalizing the interface of a wearable system to reflect the unique personality of the user.
Intimate Body Space	Space up to five inches beyond the body that the body takes into account when moving its limbs. Devices extending beyond this space can become obtrusive.
Intrabody Communication	Sending data from one device to another through the body using its natural conductivity.
Intra-Body Network	A BAN using intra-body communication.

Laws Of Wearable Systems	Five laws modeled after the Laws of Robotics by Isaac Asimov. The laws of wearable systems seek to define constraints on the actions of semi and fully autonomous wearable systems.
Learning Operational Inertia	The Operational Inertia experienced by the user while learning the operation of a device, service, or system.
Location Beacon	A device that transmits information that enables other devices to infer their location or proximity to a location.
Location Tracking	Continuously tracking the user's movements.
Lombard Effect	A person's tendency to increase their overall vocal intensity (pitch, formant location and bandwidth, etc.) in the presence of ambient noise.
Mainstream Wearable System	A wearable system that possesses the functions and ease of use that make it acceptable to the mainstream population for use in assisting them with their everyday tasks.
Mental Model	The model a user has of a device, service, or system's operation due to past experiences, world knowledge, and familiarity with the device, service, or system.
Metagrammar	A grammar that specifies the structure of commands, queries, etc. other grammars will use when defining actual commands, queries, etc. A metagrammar typically defines very few actual commands or terms itself.

Metaphoric Gestures	Gestures represent abstract depictions of non-physical form. An example is rotating your hand at the wrist as a sign to speed something up.
MicroClient	A device that contains no user applications and performs only a single simple function. MicroClients are meant to be very inexpensive and rely on the Wearable System Controller to provide the processing required to prepare the data for use by the device. An example of a microClient is a device that simply renders pixels.
Modular System	A system consisting of a core device and peripherals. The peripherals physically attach to the core device at specific docking points. The device accepts different pieces that attach to it that give it new capabilities, form factors, and interfaces.
Moore's Law	The empirical observation that the complexity of integrated circuits, with respect to minimum component cost, doubles every 24 months. It is attributed to Gordon E. Moore, a co-founder of Intel.
Multimodal User Interface	User interface that allows multiple mechanisms of interaction. An example is an interface that supports speech and GUI. Multimodal UIs usually allow the user to switch interface modes at any time in a seamless fashion. They are essential for wearable systems.
Multiple Collaborating Interfaces	A UI interaction model, in which the user can employ multiple UI mechanisms simultaneously in a collaborative manner to provide input to the wearable system.

NZOID	A Near Zero Operational Inertia Device. That is, a device which requires almost no **Setup Effort** to make it ready for use, has little or no **Interaction Complexity** as it is being used for its intended purpose, and when it is not being used, had little or no **Obtrusiveness**.
Non-Use Obtrusiveness	A measure of how often a device gets in the user's way, impedes a performance of their tasks, constrains the user's movement, or simply reminds the user of its presence when it is not needed. Non-Use Obtrusiveness is an element of **Operational Inertia**.
OI Design Audit	Identifies each application of the design principles and looks for areas in which additional application of the principles is possible.
Operational Inertia	The measure of the resistance a device imposes against its use due to its design. Operational Inertia is made up of 3 elements: 1) **Setup Effort** – the amount of effort expended in order to get the device ready for its intended use; 2) **Interaction Complexity** – the difficulty in using the device for its intended purpose, and 3) **Non-Use Obtrusiveness** – when not being used, how often the device reminds the user of its presence. The goal of wearable system design is to create devices and the system as a whole that have little or no Operational Inertia. Such devices are called **NZOIDs** (Non-Zero Operational Inertia Devices) and **ZOIDs** (Zero Operational Inertia Devices).

Opportunistic Communication	Communication that is done simply because the effort to do so is so small. Opportunistic Communication has the potential to significantly increase a person's frequency of remote communication. However, it requires wearable capabilities such as **Contextual Activation** and **Unconscious Use**.
Opportunistic Device Use	Associating with a device and using it to help perform a task and then releasing it from the wearable PAN. This is done without the intervention of the user and is typically not planned in advance.
Output Information Density	The amount of information within the output that is relevant to the user's primary task. Output Information Density should be minimized to minimize distracting the user from the primary task.
Person Area Network	A network with a range that principally covers the user's **Personal Operating Space** or beyond, up to ~30 meters. A PAN is most often associated with Bluetooth.
Personal Operating Space	A sphere centered on the person, extending no more than 10 meters in all directions, and moves with the person. The sphere defines the extent of typical interactions of the wearable system with its environment.
Personality	The complex of characteristics that distinguishes an individual or a nation or group; especially: the totality of an individual's behavioral and emotional characteristics.

Pervasive Computing	Also called Ubiquitous Computing. The concept, first espoused by Mark Weiser at Xerox PARC, envisions computing embedded in everything around us, from PCs to pieces of paper.
Pervasive Computing Environment	An environment which has computation and communication capability built into many or most of the devices in the environment.
Plan Recognition	Recognizing the plan a user has that is determining the actions he is taking.
Primary Task	A task of direct relevance to the goal of the user's activities and interest, the "real" task.
Principle Of Least Astonishment	Things should work as the user expects based upon the user's world knowledge and experience, and experience with similar devices.
Prosody	Speaking characteristics such as pitch, speed and intonation which add inflection and conveys much of the meaning in verbal communication.
Propositional Gestures	Gestures indicating places in the space around the user. They are often used to illustrate sizes or movement. An example is spreading your hands apart to indicate the size of an object.
Proximity Detection	Detecting when a device comes within a specific distance of a location.

Sensor Fusion	See **Data Fusion**.
Sensory Assistive Device	Devices aiding the deaf or blind to perceive the world around them using tactile stimulation to translate visual or audio information to touch.
Separate Simultaneous Interfaces	An application can use multiple user interface mechanisms, for example speech and GUI, at the same time. However, there is no correlation or collaboration among the interfaces.
Separate User Interfaces	Multiple user interfaces are supported by the wearable system. However, each UI is used separately and for a disjoint set of applications. There is no switching between interfaces within an application.
Service Operational Inertia	**Operational Inertia** possessed by a wearable system service or application.
Setup Effort	The amount of effort the user must expend to get a device ready for its intended use. Setup Effort is one of the elements of **Operational Inertia**. With rare exceptions, the user is not interested in the device setup and the effort to do so is seen as a necessary evil. The elimination of Setup Effort for mainstream wearable systems is a prerequisite for **Unconscious Use** and **ZOIDs**.
Situation Awareness	See **Context Awareness**.

Situational Disability	A temporary inability to effectively employ one or more of our senses or capabilities because of our current situation. For example, if the ambient noise is very high, we may not be able to use speech, rendering us effectively mute. Situational disabilities is important in wearable systems since it removes options for interacting with the system devices and reinforces the importance of multimodal user interfaces and unobtrusive devices. These are issues that can make wearable devices attractive to the disabled community as well.
Smart Clothing	See **E-Clothing**.
Social Conventions	Conventions embodying a common consensus of how we are supposed to act in public (and in some cases, private) situations. They define accepted rules of social interaction.
Sonification	The use of sound to convey information. Sonification can reduce the need for visual representation of data. This can expand the applicability of audio interfaces that typically require less use of the hands and can be less obtrusive. One area where sonification can be used effectively is in data trend analysis. By correlating the pitch, rate of change of pitch, and change in volume, it is possible to convey simple data trends.
Speaker Identification	Identifying the user by their voice characteristics. Current systems require a user to repeat back a randomly chosen phrase known to the system.

Speech Formation	Forming a message or concept into language suitable for verbal rendering.
Speech Enabled	An application or service that provides at least some of its input and/or output using speech.
Speech Generation	Rendering text as speech using a speech synthesizer.
Speech Recognition	Converting received speech into its textual representation.
Speech Synthesis	Rendering text into speech.
Speech Understanding	Understanding the semantics of received speech. This includes resolving the inherent ambiguities in a person's speech, including resolving pronoun references, and word senses.
Speech User Interface	A user interface that uses speech for its input and output.
Support Task	A task of limited user interest, typically a system or device task but with visible user benefit and primary task relevance.

System Operational Inertia	Operational Inertia experienced by a system as a whole. It is made up of the operational inertia of each of the components and that arising from the interaction among the components in the system.
Tactor	A tactile stimulator. Tactors include vibrators, force feedback generators, and percussion generators.
Technology Shedding	The act of removing most or all electronic devices from the person. Many people remove most or all of their electronic devices (among other things such as jewelry) when they come home from work.
Text To Speech	See **Speech Synthesis**.
Thermoelectric	Thermoelectric sources create energy from a temperature gradient across an interface.
Transparent Use Design Principles	Design principles aimed at eliminating sources of **Operational Inertia** and making devices, services, and systems transparent to use.
Turn Taking	The dynamic process within a dialog between two or more parties where the role of speaker and listener(s) changes from one party to another.

Twiddler	A device worn on one hand with several buttons. The user uses the buttons as a chording keyboard (using single and multiple, simultaneous button presses) to enter information. It also contains the small joystick as the mouse and uses text input buttons as the selection buttons. See http://www.handykey.com/site/twiddler2.html.
Ubiquitous Computing	See **Pervasive Computing**.
Unification	A process that determines the mutual compatibility of input from two interface mechanisms, and if they are consistent, combines them into a single result.
Virtual Speech Interface (VSI)	A mechanism to provide control of a device by speech even if the device had no speech recognizer resident. The device contains a small VSI client and a grammar that provides the speech commands for its operation. This grammar is sent to the **Wearable System Controller** when the device becomes part of the wearable system. Once the grammar is added to the speech recognizer, the device can be controlled by speech.
Visible Complexity	The complexity of a device or application that is seen by the user. Many electronic devices, including wearable systems, are inherently complex. However, if the user does not experience this inherent complexity, then it has no effect on the device's ease of use. It is only the complexity experienced by the user that diminishes the device's ease of use.

Wearable Computer	A computer that can be worn on the body. Most wearable computers are full functional replacements for the desktop PC. This renders them quite uncomfortable and obtrusive.
Wearable System Controller	The device in a wearable system that coordinates the interaction of the other devices on the system. It also contains those applications the user wants with him at all times.
Wearables	A class of devices that can be worn on the body or attached to clothing.
WIMP	The GUI paradigm employing Windows, Icons, Menus, and Pointers.
ZOID	Zero Operational Inertia Device requires no **Setup Effort** to make it ready for use, has no **Interaction Complexity** as it is being used for its intended purpose, and when it is not being used, has no **Obtrusiveness**.

INDEX

Printed in the United States of America.